Ten Nodes of Symbolizing Humanity

From the First Hominin to the Neolithic

Michael A. Susko

AllrOneofUs Publishing
Baltimore, MD & Huntsville, AL

TEN NODES OF SYMBOLIZING HUMANITY: FROM THE FIRST HOMININ TO THE NEOLITHIC

First edition. September 6, 2024.

Written by Michael A. Susko.

For the symbolic breakthroughs of early humanity.

Preface

This work has identified a core human capacity as the ability to understand and use symbols. Symbols are so interwoven in our lives; we might give them little attention. We wear clothes and certain ones, to which we may add jewelry, tattoos, and "make up," all of which have symbolic resonance. We use language; we look at images all day. We read the meaning of human faces and relate to other people's symbolic worlds. Upon reflection, we may well conclude that: "Symbolic thinking is the quintessential human cognitive ability."[1]

We do not intend to write a treatise on the meaning of symbols. We will largely leave that for the reader to infer. Nonetheless, we highlight three areas which composed the dynamic of change found in prior biological evolution. First, we noticed evolutionary leaps were accompanied by gains in *consciousness*, witnessed by enhanced sensation and an integrative function, such as the evolving neural domain in animals. Second, greater and more varied *mobility* was often paired with greater consciousness. Third, more intense sociality and *social bonding* were also associated with gains in consciousness and mobility. The primate which led to symbolizing humans was rich in each of these three categories. The order is known for encephalization, its variety of movement in the three-dimensional arboreal world, and the intensity of its bonding in social groups.

In a similar fashion, we see this threesome applied to the human symbolic dimension. An academic article highlighting the first use of red ochre half a million years ago helped to alert me to this.

> Group ritual, with its formal characteristics of amplified, stereotypical, redundant display, established symbolic culture.[2]

The practice of ritual is, in essence, a type of symbolic activity. For example, one can circumambulate a space as part of a ritual. By doing this, one is symbolically encompassing or embracing a space considered special or sacred. A person doing this is invited to enter his or her "symbolic mind." Second, you generally perform rituals through movement. It is hard to envision ritual without movement, although stillness can be a part of ritual. Additionally, the original symbolic vision may have been obtained in stillness, such as in a dream or day vision. While the original vision may be experienced as a rush of spirit, the institutionalization of the vision into ritual makes paradoxical use of repeated, redundant methods. Ritual helps bring people into a different space than everyday practical consciousness. Last, we note that ritual practice is group-oriented and bonding. Typically, we do it together, often inter-generationally, in which all ages participate.

In these nodes of "symbolizing humanity," the actual practice is often obscure and inferred, only partially grasped. Yet ritual practice is our best clue as to the prehistory of the human self and soul. Thus, we invite you on this journey to see if you can discover the roots and the future of our symbolizing capacity. Along the way we noticed these changes follow a regular, accelerating pace, which makes the journey all the more intriguing.

Introduction

The discovery of the logarithmic pacing of evolution, explored in the *Ten Pulses of Evolution*, showed the accelerating pattern of events from the start of life to the first human.[3] This chart is incomplete, however, for humans still evolved after the advent of the first hominin. Thus, this work continues the logarithmic scaling over the past 8 million years to the near present to see how the evidence sorts out.

Let's pause and consider the fancy word logarithm. In our usage, it means more things are happening quicker, that there is an accelerating pace to events. Why many contemporary people, including those in the scientific community, fail to grasp the significance of logarithmic reality would be worthy of an essay in itself. It's as natural as a tree, which starts with a single trunk, splits into a few branches, then many more branches, which multiples into twigs and leaves, each step representing a sudden jump in numbers. Or, you could take your arm, which starts with a single long bone and ends in fingers. That more is happening the closer we get to the present, or to the spatial end of something, can be intuited and examined from the evidence. In the case of paleontology, the record has become increasingly complete in the past several decades, and makes such a chart possible. We will use powers of two to create the chart, starting with the somewhat awkward fraction of two to 21.9 power ($2^{22.9}$) as it tended to fit the evidence better and produced our initial start number of 4 Gya for the beginning of life.

A major observation becomes evident as we continue a temporal charting with logarithms, from the first hominin to the start of major civilizations. That is, the evidence matches the acceleration, such we can readily show significant events at each node. This suggests the ever-enhancing capacity to use symbols, which continues in human evolution, is matching the previous rate of biological change.

The Basic Principle of Halving Units

Let us now review how the numbers on the chart are determined, keeping in mind that one does not have to follow all these details to understand the principle. Basically, we will keep halving nodes as we come to the present. This means that if we start at 7.8 Mya, the next is 3.9 Mya, the following is 1.95 Mya, and so on. In order to measure how close we are in logarithmic fashion to the Node, we will get help from by using fractions of powers of two. Thus, 22.9 raised to the power ($2^{22.9}$) gives us the first human node at 7.8 Mya. If we go one power less, or 2 to the 21.9 power ($2^{21.9}$) this gives us half the value, 3.9 Mya, and so on.

In gaining precision to map specific events, we use 10%, 15%, 33.33... % log distances from the Major Nodes. In the case of 10% difference, the node at 2 to the 22.9 power ($2^{22.9}$), which gives us the value of 7.8, becomes $1/10^{th}$ less or 2 to the 22.8 power ($2^{22.8}$), producing the value of 7.3 Mya. Having such expanded and detailed mapping gives us the resolution we need to map the continuing, accelerating pace of evolutionary events.

Again, don't let logarithmic numbers confuse or detain you. The end result is that we are using a chart which comfortably fits the accelerating nature of change. This forgoes plodding along in a linear fashion and seeing what happens at 10, 9, 8, 7, 6... millions of years ago, creating considerable empty spaces. Rather, we will move from 10, to 5, to 2.5, to 1.25, to .625 Mya, and so on...obtaining a finer resolution as we come to the present—one which successfully captures more frequently occurring events. As noted earlier, the significant nodes of change with humans, which synergizes with their symbolizing capacity, continue the previous biological pacing.

This work covered *ten* nodes of symbolizing humanity. I considered doubling this number, covering the logarithmic midpoints.

That is, we could subtract .5 from a power and see what happens in the middle of the nodes. At times we found developments which suggest the normal logarithmic rate of change may have doubled. With the introduction of a "symbolizing capacity" in humans, this would not be surprising. However, twenty nodes risked introducing more complexity than needed, and one that might obscure the main events. As it is, we occasionally refer to mid-nodes when it helps to further our narrative and makes it more complete.

Any tool overlaid on the evidence is like a lens, which should help us to see things clearer. A "logarithmic lens" is a good all-purpose lens, providing a wide angle, a mid-range, and a macro or detailed view, to view the nature of change. We are almost ready to view the evidence in a novel temporal way, one which captures main events, includes considerable detail, and makes clear the accelerating nature of evolution.

The Precursor Period, the Major Node and the Explicit Period

The simplicity of halving of units and looking for change clustered around Major Nodes serves as a powerful tool. We can easily remember halved units, from which we can map data around to see how evidence sorts out. This hides a complexity, however, the framing of logarithmic distances around the nodes. Logarithmic spacing means we are expanding the spatial mapping, so we have more resolution as we come to the present. Thus, one-fourth or 25% of the log distance prior to each node is a longer linear number than one-fourth log distance *after* the Major Node. For example, with our first node, 25% log distance prior to the Major Node is 1.5 Mya, while the same 25% log distance after the Major Node is 1.2 Mya. That means the same log distance in the second measure is 300,000 years or 20% less.

Next, we consider the usage of the words Precursor Period, Major Node and Explicit Period. Our expectation is that the Precursor Period, log distances prior to the Major Node, will have events which are pointing toward a major change. The Explicit Period provides confirmation that such change has definitely occurred. This means the Major Node occupies the logarithmic middle, and which serves as the point in time in which the change is hypothesized to have occurred. This structure accounts for the reality that we are highly unlikely to find evidence at the actual moment of change. We will generally mark 10%, 25% and 33% distances on either side of the node. We consider the 10% marker s almost a direct hit to the Major Node with one-third stretching the outer boundary.

This disparity between evidence and actuality of when events first happened has not escaped notice by scholars. In one formulation, the time of the earliest fossil find is labeled the "Biozone," with the actual projection of the start of the change being the "Chronozone." In-between, a missing period of evidence is called the "Range Offset." The mathematics of predicting the actual first occurrence is complex and depends on determining the abundance of a species, and having enough finds to determine its rarity. Hominid finds are notoriously rare, so it is hard to get a proper sample.[4] While this mathematics sorts out, and which in all likelihood will remain an informed guess, we propose a logarithmic mapping to order the evidence. Thus, this model proposes a constant rate of acceleration, with regular nodes highlighting significant change and for which evidence can confirm.

We are just about ready to venture into the actual data of events and test our model's ability to effectively narrate our evolutionary story. But first we will identify a pattern among a series of Major Nodes.

Patterning with a Series of Four Major Nodes

From my earlier logarithmic study of biological evolution, a pattern with a series of four Major Nodes was identified that helped organize the flow of data. Typically, in all the nodes of human evolution, change is associated with significant developments in consciousness/ brain development, mobility, which relates to bipedality and concomitant expansion of range, and sociality or capacity for cooperation. Yet within this general frame, a further pattern was revealed as Major Nodes unfolded.

1) First, a quantum change, larger than normal, starts the series. In this work, the first quantum change is the start of humanity with its beginning sense of consciousness. The first node in a series has the potential to generate further major change. Invisibility often surrounds the first node in a series. While an array of evidence indicates such change, it is more uncertain than the following nodes.

2). By the second node in the series, we have more sure confirmation of the evolutionary leap, with scholars more readily agreeing the change offered in the prior node has indeed occurred. With this node, we are particularly attentive to a novel social fusion and a geographic spread out of the initial inhabited zone.

3). By the third node in the series, we find further elaboration and grade level change in the three dimensions. In particular, signs of increased neural capacity with attendant gains in consciousness become more clearly manifest. Lastly, we sense a gathering potential for the next quantum level change.

4). The fourth doubles as another first in the series, presenting the next quantum change.

This general frame is still being articulated and explored. It serves as an invitation to consider the possibility of how leaps in evolution are patterned. A presentation of this pattern can be illustrated on a type of zigzag map, which we will place in the appendix.

We are finally ready to present the evidence for the ten symbolizing nodes of humanity.

Chapter I

1st Hominin (Human) Node: Consciousness

7.8 Mya ($2^{22.9}$) 10th Node of Life

This first node of change is also the tenth and final node, presented in the first volume, *Ten Pulses of Evolution,* which traced the arc of life from its start to the first human. Here we briefly review and expound further upon this node, for it sets the stage for symbolizing humanity

With the first human, we posit the presence of consciousness, a radical step in the evolutionary landscape of the universe. With consciousness, the capacity for a symbolizing human more squarely emerges. The essential physical changes which can be documented and which point to this capacity are habitual bipedalism, qualitative changes in head shape and size, and a reduction in canine size.

These three areas of change provide the first evidence for "symbolizing humanity" at this node. Rather than direct evidence like tools or cultural artifacts, the bones of human bodies in an archaeological context arguably show the attainment of human status.

To review and expand, three traits were emphasized and documented: 1) Habitual bipedal and upright behavior, which favored freedom of movement with arms and hands to do an ever-increasing variety of tasks. 2) A head and face, which in males become less aggressive appearing, due to diminished canines more equivalent to females, and behaviors inferred from greater pair bonding between male and female. And 3) a more human-shaped head with double the brain size, an increase that kept pace with the increase in body size. A paired increase in body and brain size should not be underestimated in terms of evolutionary significance. It meant a greater capacity to act in the world and more space in the "organizing center" or brain, to harbor electrical fields and their interweaving.

With these three traits in mind, let us see if we can go a step further and ask, what might they mean in terms of a "symbolizing human"? A human upright to the world has his or her face more exposed and the full length of the frontal body. The face on this side of the body can be seen as more expressive. To some extent, this "underside" also projects vulnerability, one which could make for more intimacy for social bonding.

At the same time, height also conveys strength and a more dominant position in the world. Instead of a crouched quadruped with flaring canines, humans, who shed their long canines, stood upright. The beginnings of upright postures goes back to some extent to the great apes. Hominins, or early humans, however, increased their ability to project into the world by their habitual upright posture, and habitual bipedal journeying/walking together.

This upright/bipedal way favored the expansive use of arms and hands, and the subtle expression of gesturing. As this type of primate ultimately became the more expansive, symbolizing one, we conjecture it is related to their mode of movement and expression in the world. In this vein, many theories have focused on gesturing as the basis for the emergence of symbolic communication and language. [5]

From the above, and at the risk of repetition, we posit that symbolic capability ultimately started with the embodied human, which interacts with the world in a physical way. As one researcher put it, "Language derived from bodily movement."[6] Before humanity spoke using symbolic words, the body, with all its expressivity, communicated. If we follow recent discoveries regarding the amazing fluidity of the genome, and the inheritance of electrical fields, we might go so far as to say that humans who adopted a psychology of "upright and journeying together," were the ones who evolved and became human. In this vein becoming human was a choice, and one we are still making today.

With the advent of humans, we are ready to begin a detailed analysis of logarithmic patterning of events and their relation to a symbolizing capacity, which will continue for nine more nodes.

Chapter II

2nd Hominin Node; Kenyanthropus & Australopithecines

3.9 Mya ($2^{21.9}$) 11th Node of Life

The second node of a series is marked by a clearer establishment of the hominin type, novel social fusion, along with marked behavioral/ neural change. Typically, there is also expansion into a new ecological zone. By the second hominin node, we find clearer evidence for more *habitual* bipedal behavior, neural enhancement, and ecological expansion into open habitats, as well as signs of increased bonding and cooperation.

In the region of the mid-node (5.53 Mya), *Ardipithecus kadabba* is dated 5.77 - 5.54 million years ago. This is preceded by a molar, tentatively assigned to this species and which has been dated to 6.3 Mya, at the prior 1/3 explicit node.[7] Haile-Selassie has concluded that a critical toe bone supports the presence of a "well-adapted biped." [8] Although debate remains about its relationship to later hominins, Hall and Hallgrimsson believe the teeth and postcranial traits of *A. kadabba* indicate a clear hominin. Importantly, the teeth from *A. kadabba* reveal their canines have lost the honing capability found in great apes, meaning their canines are no longer routinely sharpened by rubbing against one another.[9] According to anthropologists, a reduction in canine size means these hominins lived in a "radically less competitive social structure than chimps."[10] Further, it's hypothesized the innovation of mother-centered "gathering" involved a group of food-sharing females rather than the nuclear family.[11] We can also envision males becoming more involved in the parenting process, carrying infants and food, which also worked to accelerate the bipedal trend.[12] These various hypotheses are consistent with our expectations for a more intense social bonding at this node.

Between the 1/4 and 1/10 precursor nodes, *Ardipithecus ramidus* has been dated from 4.5 to 4.3 Mya. *A. Ramidus* inhabited a mosaic of habitats, from moderate rainfall woodlands to woodlands/ grasslands.[13] Its fossils present a mix. Their teeth have thin enamel like the great apes; their arm bones possess both hominin and ape-like traits, which suggests they did not bear weight; and other traits indicate bipedality. According to Rice and Moloney, the position of the foramen magnum (the hole in the base of the skull) makes it "very likely" that *A. ramidus* was bipedal.[14] Although their canines had functional honing, they were lower, blunter, and more incisiform than those of apes, thus exhibiting a hominin-like reduction in sexual dimorphism.[15] Body sizes between male and female specimens were also close to one another. Begun believes that *Ardipithecus ramidus* is the only taxon with clear, unambiguous similarities to later hominins.[16]

The recent reconstruction of a nearly full skeleton of *Ardipithecus ramidus*, dated 4.4 Mya, provides us with a confirmation of earlier finds, as well as surprises. The 110-pound female specimen possessed an unexpected mix of advanced traits and primitive features found in much older apes than chimpanzees. This hominin had its own unique locomotion, moving bipedally on the ground and quadrupedally in trees. Using her palms for movement in trees, she retained flexible wrists, unlike contemporary great apes. This mixed form of locomotion, along with smaller post-canine teeth with thinner enamel, indicates "significant behavioral and dietary change."[17] Last, we note her stubby, diamond-shaped canines were "radically feminized," indicating movement toward hominin status.[18]

A little after the 10% precursor node (4.17- 4.02 Mya), *Australopithecus anamensis* lived in the woodlands of South Africa, which Pickford marks as the arrival of "fully bipedal hominids."[19] White and others have placed *A. anamensis* as temporally and anatomically intermediate between the prior *A. ramidus* and the later australopithecines. Over a 200,000-year period, between 4.4 and 4.2 Mya, a "burst of rapid directional change" is posited.[20]

This early Australopithecine appears to be the first to enter into more open Pliocene habitats, in what has been termed an "ecological breakout."[21] Consistent with this portrait, *A. anamensis* possessed thick-enameled teeth, indicating a diet of hard to chew foods.[22] The canine tooth crown height, estimated from their roots, is similar to that of modern humans, "suggesting a low degree of sexual dimorphism." [23] A partial tibia, dated 4.2 - 3.5 Mya, has traits which indicate bipedalism, including a thickening of bone where it meets the ankle, which serves as a shock absorber. Last, from 4 to 2 Mya, *A. anamensis's* brain increased in size from 400 to 510 cc, a 25% increase. Although the traits of bipedality, thicker tooth enamel, reduction in sexual dimorphism, and an encephalizing trend harmonize with the human lineage, *A. anamensis's* place on the phylogenetic tree is unresolved.

With the introduction of the Australopithecines, it is important to evaluate their general status in regard to hominins or humans at this second node. Historically, this group has been seen as intermediate between a semi-erect quadrupedal (on all fours), chimpanzee-like ancestor and obligate bipedal humans with a large brain. However, this has been increasingly challenged with a counter-hypothesis that this group are ancestors of the contemporary apes, who later developed knuckle walking.[24] In this view, rather than hypothesizing a knuckle-walking ancestor, a more erect bipedal orientation is seen as having already developed, as indicated by our review in the prior node. This uncertainty as to the next step needs recognition so we can be open to new evidence. Doubts were raised decades earlier in which researchers found that many of the Australopithecine's traits fit ape evolution better than human. Partly the reason is that researchers want to extol the human connection to their finds. Another problem is that reconstructions of the ancestral or primitive hominoid condition may fail to grasp that humans may be more "primitive" in some traits than the apes. Last, it's important to keep in mind the australopithecines are not a definite genus, rather, they are a loose assortment, a "collection of experiments," who can be seen to "clearly represent the sister group to Homo."[25]

The paucity of great ape finds is remarked upon in the literature, but maybe great apes have been hiding in plain sight under the Australopithecine label. When I read a current evolutionary text describing Australopithecines, I cannot help but notice how often their similarity to apes is described. These include the critical ratio of proportion of top of the arm bones (humerus) versus lower arm bones, radius and ulna: hand bones curved like modern apes, short thumbs, sexually dimorphic in body size, and a rapid rate of maturation like chimpanzees.[26] Last, a current review admits, "major questions are unanswered, that both the origin of Australopithecines is unknown, and the MRCA (most recent common ancestor of humans) "continues to elude researchers."[27]

This problem can be framed more positively from the primate perspective. The primate Bauplan is one capable of producing various patterns of mobility in the rich and variegated three-dimensional world of trees. It is this evolutionary fertility which has left us guessing as to which one definitely led to humans.

The region around the Major Node presents us with classic finds of *Australopithecus afarensis.* Whether this genus was on the line to humans or great apes, her finds can be seen as representing the evolutionary potential present at this node. The earliest finds for *afarensis* are teeth that have been dated to 4.2 Mya, about 10% prior to the Major Node. The more confirmed finds for the species appear at 3.6 Mya, 10% after the Major Node.

Being right in the range of the Major Node invites us to give this find particular attention. First, their cranial capacity size varied between 375 and 550 cc, up to a third larger than the *Sahelanthropus* cranium from the previous node. Second, *A. afarensis* displays a more generalized dentation, suggesting that they traveled distances to access varied foods. Third, the post-cranial skeleton of *A. afarensis* demonstrates a capacity for regular movement on the ground. Supporting evidence includes an elongated neck in the thigh bone, wide and short hips, and the angle of articulation in the knee joint and feet. A number of traits, however, were useful for climbing: good upper body strength, a lower center of gravity for stability, a shoulder joint useful for considerable overhead activity, and hands and feet adapted for powerful grasping.[28] In short, our impression of the australopithecine body plan is that this hybrid being could manage the arboreal environment as well as moderate land activity.

As already noted, anthropologists generally interpret a reduction in sexual dimorphism to mean male and female were on a more equal footing, a state associated with monogamy and other modern forms of human behavior.[29] Traditionally, *A. afarensis* has been characterized as strongly sexually dimorphic, but a reevaluation posits a relative similarity between the sexes in "skeletal dimorphism" and "feminized male canines." [30] Some scholars hold the canine difference is "much less pronounced" than in apes, or early Miocene pongid fossils from two nodes ago. Yet, the canines are about the same relative size as extant bonobos, with size differences between the sexes seen as not substantially greater than modern humans.[31] A recent study, however, describes a mixed pattern in terms of limb dimensions. When looking at the leg's femur, Lee found that *A. afarensis* was similar to gorillas, who have a large degree of sexual dimorphism. On the other "hand," the arm's humerus was found to be more similar to equally sized humans.[32] In sum, *A. afarensis* presents a mixed picture with regard to a reduction in sexual dimorphism, a basal trait of early hominins.[33] Although *A. afarensis* is the most completely studied hominid, questions remain, as we might suspect, whether it belongs on the line leading to humans or is a sister clade.

We next turn to *Australopithecus prometheus*, the most complete find of the "genus," dated 3.67 Mya, at the 10% explicit node. This species was competent on land, able to walk "medium distances," and effective in trees, exhibiting "locomotor plasticity." An affinity with the great apes is supported by several specialists' assessments that their "postcranial features seem to be built on proportions that are primitive for living great apes as a whole."[34]

If not the Australopithecines, we are left asking which group might serve as a "missing link" to the *Homo* family? We turn to a distinct hominine, who lived at the same time as our prior example, revealing the deep diversity of the hominine group.[35] Between the 1/10 and 1/4 explicit nodes (3.65 and 3.3 Mya respectively), our hope for a clarifying find is realized with *Kenyanthropus platyops*, which has been dated 3.57-3.4 Mya.[36] The skull presents a flatter/human-like, facial morphology and smaller cheek teeth, traits not generally seen in the record until a million years later.[37]When considering adaptive zones in the definition of genus, *K. platyops* can be considered a "species germinalis," an early species of the new *Homo* genus.[38]

During the explicit period, a critical development occurred in the symbolic evolution of humanity. At 25% log distance after the Major Node (3.3 Mya) stone tools labeled LOM3 were found in West Turkana, Kenya, a million years earlier than any tools associated with the genus *Homo*. These relatively crude-looking, single-edge "choppers"—or the simple flakes they produced—don't look like the revolution they represent. In what has been termed the "bipolar method," a source stone is identified and struck with a hammerstone—likely a rounded, harder stone—producing a sharp flake for use.[39] The conceptual sequence bears repeating: a source or "mother" stone—struck by a "father" hammerstone—produces an "offspring" flake for tool use. Importantly, the source stone is also placed on a stable stone platform before striking, thus the word "bipolar." In sum, four stones are involved—a quaternity—further enriching the process and its symbolic resonance.

Later developments will make this process clearer, and the more refined results will capture attention and be credited with the tool revolution. Yet the evidence shows we can trace this process as early as the second node of symbolizing humanity. That is, at least for the process involving stone. Wooden tools, or even a woven nest—using a similar conceptual frame—likely preceded this.

From the results we can also deduce physical evolutionary changes. The LOM3 artifacts indicate substantial hand and motor control, and cerebral cortex elaboration, especially the motor cortex.[40] Relatedly, we also ask what group made these tools? At this time the only proposed hominin group located in this area of West Turkana, Kenya, was *Kenyanthropus platyops*, whom we just considered. If I were a detective on the hunt for human origins, *K. platyops* is emerging as a prime suspect.

Meanwhile, more evidence presents for *A. afarensis* at the 1/4 explicit node. Dated to 3.3 Mya, Salem presents the most complete skull and skeleton of a juvenile *A. afarensis.* Study shows that she walked, even at the age of three, and possessed a hyoid bone, allowing for production of chimp-like sounds.[41]

At the 1/4 explicit node (3.3 Mya), *Australopithecus africanus* also appears, which Cameron believes is the best candidate for the earliest confirmed hominin. *A. africanus* has smaller male canines, larger and thicker enameled molars, and was adapted to some form of bipedal locomotion. Also, the species presents an absolute increase in encephalization, an absolute increase in molar enamel thickness, and an absolute reduction in male canine size.[42]

In terms of sexual dimorphism, *A. africanus* had a moderate to high degree of shape dimorphism in its skeleton and a moderate degree of size dimorphism, somewhere between the chimpanzees and the orangutan.[43] *A. africanus* has human-like proportions in terms of its thumbs' length, although it is more ape-like in breadth.[44] However these limb and hand proportions sort out, *A. africanus* possessed a wide range of mobility as it expanded into dryer, more seasonal environments. The presence of C4 in *africanus's* teeth verifies that they ate tropical grasses and sedges or animals that ate these foods.[45]

Whether *A. africanus* is on the human line, evidence is suggesting grade-level change. A recent reevaluation of the Encephalization Quotient (which factors body size with brain size) found that the EQ of *A. africanus* is comparable to *Homo* of the next node. In addition, the degree of basicranial flexion (the angle between skull and body) is consistent with having the throat structure for language, making *A. africanus* similar to much later humans.[46] Later skulls exhibit novel brain indentations in the region of the posterior parietal cerebral cortex, believed to be involved in the multi-modal processing related to social behavior and an expanded ecological range.[47] Last, with hearing, *A. africanus* displays a gain in auditory capacity for the frequencies in which language is heard. Peak energy gains in the area of 2.5 kHz are somewhat heightened when compared to chimpanzee's whose range is between 2.5 and 5 kHz.[48] In sum, varied capabilities serve to give *A. africanus* candidate status for being on the human line or a sister clade to the human line.

At the 1/3 explicit node (3.1 Mya), we come to the most complete early adult hominine skeletons, which have served as a reference point for other fossils. "Lucy, from Hadar in Ethiopia, had a small, ape-sized brain, with pelvis and leg bones almost identical to modern humans. Also dating to this time and from Hadar, the most complete skull of *A. afarensis* has been found, as well as the most complete mandible of any Australopithecine.[49]

Last, at the 1/3 explicit node around 3 Mya, we call attention to a red-brown cobble made of jasperite, found with human remains. This half-pound stone has unusual natural markings suggestive of a face. As the cobble was smoothed by water, yet found a distance away in a cave, it was likely a manuport carried by humans.[50]Bednarik characterizes this as the earliest example of an "exogram," showing the human ability to connect to a symbol, or to "create a construct of reality." This sensing of meaning from an external object, which is also aesthetic, prefigures modified stone objects which arise later. [51]

Let us now review the evidence in chart form:

*1/3 Explicit node (3.1 Mya) Lucy skeleton 3.1 "Face-like" Jasperite, South Africa cobble 3

Paranthropus 2.9 -1.2 A. Afarensis 444.2 complete skull 3 Mya bipedal ape

*1/4 Explicit node (3.3 Mya) *Australopithecus africanus* 3.3

Stone tools, West Turkana, Kenya 3.3Mya

Many named species of Australopithecus 3.6-3.2

*10% Log distance (3.65 Mya) *Kenyanthropus platyops* 3.57-3.4 Footprints 3.59

Austalopithecus Prometheus 3.67

$2^{21.9}$ ❂ **Major Node** (3.9 Mya) *A. afarensis/competent biped /almost entirely apelike skull* 3.8-2.9 *10% Log distance (4.2 Mya) *Australopithecus anamensis* 4.2 - 3.9

*1/4 Precursor node (4.65 Mya) *Ardipithecus Ramidus* 4.5 - 4.3

*1/3 Precursor node (4.9 Mya)

+ Mid-node (5.53 Mya) *A. kadabba* 5.8 - 5.2 /Loss of canine honing

In summary, after the early find of *A. kadabba,* finds are tightly clustered within 1/4 log distance on either side of the Major Node. Generally, this node is characterized by well-established bipedal hominids with enhancements in encephalization. Evidence presents for reduced dimorphism (especially canine reduction), suggesting reduced competition between males and more equal footing between the sexes. Of the various types, *A. Africanus* has historically emerged as a contender for being on the human line. Yet, a reevaluation of the evidence asks us to consider *Kenyanthropus platyops* as a viable contender, given its human-like face and association with the earliest stone tools found. However the evidence sorts out, a number of evolutionary potentials are expressed in bodily form and the finding of stone tools, giving us justification for envisioning a grade level change at this node.

The stage is now set for the emergence of the *Homo* family. Our expectations are heightened as we near the relative center of a logarithmic scale, which started from the origin of life and takes us to the cusp of historical times.

Chapter III

3rd Hominin Node; Homo family 1.96 Mya

($2^{20.9}$) 12th Node of Life

With the third node in a series, we look for a significant elaboration and heightened social bonding, as well as unequivocal evidence for the human type. By this time, the basic human form has been achieved, with hominins becoming more resilient as reflected by size increase, geographic spread, and technological gains. As with all Major Nodes, we expect significant enhancements in the neural/cognitive sphere.

This node presents the emergence of the *Homo* family, an event marked by clear brain expansion and cultural innovation. Contrary to the expectations of early paleontologists, the more radical encephalization of humans occurred *after* the habitual bipedal behavior evolved. A movement-first scenario would open up behaviors that stimulated and favored added neural capacity. At this node, the human body also underwent an increase in size, which achieved the basic human body proportions, which enabled long-distance travel. Human beings became obligate land dwellers, no longer incorporating an arboreal dimension into their lifestyle. This may be related to an expansion of savannah which started prior to the Major Node, between 2.5 and 2.4 Mya.

Let us turn to the archeological and paleontological evidence. Between 2.6 and 2.53 Mya (between the mid-node and 1/3 precursor node) early Oldowan tools appear in Gona, Ethiopia.[52] Rounded stones were knapped (with edges knocked off), making choppers and flakes. Although some comparison can be made to great apes, who use tools on an ad hoc basis, a sophistication beyond primate tool use was involved. Oldowan toolmakers demonstrated an empirical understanding of how hard rocks fracture and would make adjustments accordingly.[53] They are also unique in making tools from other tools, as flakes were made by using a softer hammer stone to strike a harder core stone.[54] Finally, early humans carried cobbles, anticipating the need to make tools at different sites. This more conscious production of tools coincides with the emergence of the *Homo* family.

An association between stone tools and bones with cut-marks, in a good archeological context, comes from Bouri, Ethiopia. Early Oldowan sites, dated 2.58 Mya, reveal that large game animals were butchered. Sharp flakes produced from cores gave ready access to meat and marrow, high-quality food sources which abetted survival. Rather than actively hunting large animals, detailed analysis of the pattern of bones and their cutmarks indicates that early humans more likely *scavenged* from carnivore kills.[55] Rather than longer term "home bases," these sites were more likely temporary "scavenging stations," which still entailed cooperation needed to defend their sites from carnivores.[56]Such advanced scavenging suggests a planning capacity, the ability to strategize and remember past events.[57]

At 2.5 Mya, close to the 1/3 precursor node, we have finds of *Australopithecus garhi,* meaning "Southern Ape Surprise." The cranium *of A. garhi* possesses a unique and unexpected combination of primitive and new features.[58]A relatively large skull with an estimated cranial capacity of 450 cc represents an increase in brain size among the Australopithecines.[59] In the postcranial body, a mix presents, with arm proportions which are ape-like, and with legs (femoral elongation) having *Homo*-like proportions. *A. garhi* also has large teeth (megadontia) and larger canines than any *Australopithecine* or *Homo* species. In Bouri, Ethiopia, *A. garhi* has been associated with butchered meat and marrow from large mammals, which have evidence of cut marks and percussion marks clearly made by stone tools.[60] Thus, in terms of brain size, leg proportions, and possible tool use, *A. garhi* can be considered either an ancestor or a sister clade to early *Homo*.

At the 1/4 precursor node (2.33 Mya), more advanced Oldowan tools appear. Paleontologists have deduced a structured mode of production followed "constant technical rules resulting in high productivity." From Turkana, Kenya, an abundance of finds, dated to 2.34 Mya, are unique in showing this technology.[61] Tools, which required very careful preparation, demonstrate the cognitive capacity to envision and solve problems.

Fossils as old as 2.4 million years have been attributed to the *Homo* family. Right at the 1/4 precursor node (2.33 Mya), a member of the genus *Homo* (specimen A.L. 666-1) has been associated with tools. Although species attribution is difficult to assign, this fossil find appears to be the earliest representative of *H. habilis.* Most finds for early *H. habilis* fall between 1.87 and 1.66 Mya, right after the Major Node.[62] The cranial bones of *H. habilis* are relatively thin, with a braincase expanded side to side, and the face tucked more under the brain.

Consistent with this general timing, a mutation is hypothesized to have occurred at 2.4 Mya, leading to early *Homo* traits: diminished jaws, a "pattern of gracilization," and "accelerated encephalization." [63]From brain expansion, we can also infer further neural differentiation. Weaver details a shifting cerebellum and neocortex ratio, with neocortex acceleration dominating early human evolution. [64] In sum, evidence converges around the Major Node for a gracile look, marked encephalization, and sophisticated tool use.

The projected body proportions of *Homo habilis* have been subject to dispute. Early studies from fragmentary remains led to the belief that the limbs of *Homo habilis* represented a more regressed, ape-like condition from two nodes prior. A recent re-evaluation has granted *H. habilis* more modern, human-like proportions, as we expect at a later node.[65] A follow-up study by the same authors reveals a more complex picture, confirming human-like proportions in the leg (humerofemoral index), chimpanzee-like brachial proportions, and a more ape-like shoulder girdle and forearm length.[66] Pongid features are present in the flexor muscles of their fingers, indicating the tree habitat was not totally abandoned.[67] In short, *H. habilis* has a hybrid mix of locomotor capacities.

Consistent with this portrait, controversy exists over fitting early *habilis* specimens into a single species as finds are so variable. This confusion, along with its mixed features, is consistent with our expectation that the precursor period presents material difficult to categorize and is dynamically changing.

Questions have arisen as to the unique difference between *H. habilis* and the prior Australopithecines. Traits such as thicker bones in the braincase, a higher cranial vault, and a flatter lower face appear to be one of degree. On the other hand, brain size and more human-like proportions in the legs indicate that a grade-level change is underway. Whether the change is one of degree or more qualitative in nature, a case can be made that *Homo habilis* is a precursor to grade-level change.

At 2.2 Mya, about 1/10 log distance before the Major Node, novel tool-making techniques, which become more typical later, have been found in Kanjera, south of western Kenya. Centripetal flaking was employed to make a core with extensive cutting edges. Additionally, the life of cores was extended by rejuvenation as flakes were retouched. [68] These finds present an improved tool technology during the early *Homo* period.

A change in the human body, critically related to tool development occurred with the human hand. In the first study that took into account soft body parts as well as skeletal, the advent of "efficient thumb opposition" has been found to occur approximately 2 million years ago. Parts of hands, due to their likely scattered locations, are often difficult to associate with a given species. However, this study shows that these changes did not characterize Australopithecines, but other hominins, possibly early Homo, as later Homo species show "high levels of thumb opposition" dexterity.[69]

Perhaps related to this is a stunning development of finding unexpectedly early *Acheulian* tools. Dated to 1.95 Mya, right at the Major Node, an infant mandible of *Homo erectus* has been documented at a high altitude in Ethiopia. Here, tools were produced in a complex sequence of steps, in which a core was produced first, then both sides were chipped to make a bifacial tool.[70]

More complete evidence for the emergence of *Homo erectus* occurs at Koobi Fora, Kenya, dated between 1.95 and 1.9 Mya, seconding this fuller elaboration of the *Homo* family. The body size of the genus had become large, some 50% greater in body mass than *H. habilis*. Historically, *H. erectus* has been viewed as obtaining the basic modern proportions of the human body with a more linear, slender shape and a domed-shape head. Along with its longer shape and definite waist, its completely extended lower legs were capable of long strides. However, this basic view is subject to some modification. A recent 3-D reconstruction of its rib cage anatomy from a juvenile skeleton in Nariokotoma, Kenya concludes that its shape was relatively stocky compared to *Homo sapiens*.[71]

Homo erectus was clearly on the path to *H. sapiens* as an obligate land dweller who could traverse long distances. Longer legs are believed to have compensated for new energy costs associated with increases in body mass and brain metabolism, slower rates of childhood birth, and more daily movement.[72] *H. erectus* may also have been capable of *endurance running*. A variety of adaptations, including changes in the feet and a counter-balancing weight in the buttocks, have led paleontologists to conclude this uniquely derived trait emerged two million years ago.[73] In sum, *Homo erectus* was approaching human size and shape and had a modern capacity for movement.

Greater mobility proved useful for survival, given the ecological changes at this node. Several lines of evidence converge to indicate the presence of C4 grasslands, grazers, and increased aridity.[74] The shift to more open landscapes and more diverse habitats with a patchy distribution of resources placed demands on the hominin with its expanding brain.

A recent study of the earliest Olduvai hominins at 2 Mya (the Major Node) concluded they exploited unstable habitats, one which "pioneered a rapid occupation of geo-settings undergoing drastic changes" and the utilization of "disturbance environments." With their generalized tool set, early hominins showed a behavioral flexibility and adaptability, which later in this node proved able to expand globally.[75]

While movement can be emphasized, so can the centralizing capacity in establishing "home bases". Recent evidence, as with David's site in Olduvai Gorge, dated 1.84 Mya, supports the presence of "central provisioning places." Large carcasses were regularly obtained by actively hunting adult bovids, which were then transported to a central site to be processed and shared. This demonstrates that in the period from 2 Mya to 1.8 Mya, a leap occurred in human social structure and associated cognitive traits.[76] A recent summation describes it this way.

> ...*Homo erectus* had transitioned away from an apelike social system and toward a way of life more like modern human foragers—a lifestyle in which complex, knowledge-intensive foraging led to the evolution of cooperation, division of labor, and a slower life history.[77]

The advent of *Homo erectus*, with its larger brain and body size, has often been associated with an increase in meat-eating. The presence of central provisioning places for processing animal carcasses would support such a view. However, a recent study introduces a cautionary note that between 2.6 and 1.2 Mya, the zooarchaeological evidence does not support an increase in carnivory.[78] Perhaps the answer for the changes is more related to the use of enhanced technology, more cooperative social domains, and the challenging environments that early humanity faced.

Let us turn to specific fossil finds marking the transition from early *Homo* to *Homo erectus.* Around the Major Node, the KNM-ER 1813 specimen has been dated 1.9 - 1.8 Mya. Usually assigned to *H. habilis* due to its cranial shape and other traits, the specimen also has "advanced traits" associated with *H. erectus.* KNM-ER 1813 has even been assigned to *A. africanus,* due to its small brain size of 510 cc and adult developed molars.[79]Such uncertainty is consistent with our expectation for a transitional zone near the Major Node.

Between 1.9 and 1.8 Mya, the famous Skull KNM-ER 1470 has been measured to have an expanded cranial capacity of 750-775 cc. Often assigned to the *H. habilis* group, this skull has been reclassed by some paleoanthropologists as a unique species, *Homo rudolfensis,* due to measures which granted it a significantly larger brain and a flatter face. A recent reconstruction, however, has called into doubt the flatness of its face, which impacts its estimated brain size, placing it closer to 525 cc and within the *habilis* range.[80] The question remains open as to whether *H. rudolfensis* crossed the cerebral Rubicon of 600 cc, a measure traditionally marking entrance into the *Homo* family. Evidence from endocasts at least points to a reorganization of the frontal lobe and greater cerebral asymmetries, closer to the modern condition.[81]

A few hundred thousand years after Skull 1470, *Homo erectus* appeared with a body that presented more human-like proportions, although these proportions are also found in some *H. habilis* fossils. Dated 1.78 - 1.63 Mya, between the 1/10 and 1/4 explicit node, an African specimen, labeled KMN-ER 3733, possessed an expanded brain capacity at 850 cc. The brain has expanded due to larger body size, not more in proportion to the body. In all likelihood, a larger brain also underwent more differentiation, especially with areas impacting communication. Broca's and Wernicke's speech areas have been detected on surface endocasts, leading some scholars to believe the neural basis for spoken language had evolved by this time.[82]

In the region of the Major Node, evidence emerges for social and behavioral complexity. Transport of food and stone tools had become habitual.[83] Lithic resources, dated 2.3-1.95 Mya have been culled from varied places in the ancient landscape as much as 10 km distant. [84]By 1.9 Mya, social networks expanded in size, indicating a planning capacity greater than the chimpanzee's.[85] Habitat reconstructions of early *Homo* are marked by significant heterogeneity and utilization of a broad variety of settings.[86] This includes access to tougher foods than those utilized by *H. habilis*, as indicated by the micro-wear patterns on teeth.[87] In sum, *Homo* had more mobile and larger social networks that were able to access a broad range of resources.

About 1.8 Mya, hunting and/or active scavenging had become a strategy to obtain food. A macro-mammal bone assemblage from Olduvai exhibits teeth marks, which show that hominins had *early* access to fleshed carcasses.[88] This does not rule out other strategies for gaining early access, such as scaring off predators by making noise, as do the Hazda hunter-gatherers from Tanzania.[89] Although appearing at variance with the image of prehistoric man as a dominant hunter, such activities demanded substantial cooperation. *Homo erectus* was also compassionate. A 1.77-million-year-old skull without any teeth suggests that they shared food with the elderly.[90]

Tantalizing clues point toward an increasing potential for speech. A complete spinal column from the Dmanisi *Homo georgicus*, dated 1.78 Mya, has led researchers to believe that fine locomotor control for breathing made speech possible.[91] Handedness has been associated with language, and a right-handed preference for making stone tools was present by 1.7 Mya.[92] In sum, the biology of *H. erectus* presents several traits for the permissibility of language.

Other signs point toward an intensification of cultural life. From Bed 1 at Olduvai, dated 1.75 Mya, a semi-circle of basalt chunks—a possible structure—contains a large concentration of tools and bones. [93] Anvils with "artificially pecked" depressions appear to have served as nutcracking tools and possibly held symbolic meaning as "cupules." [94]

Recently, a 1.7 Mya Acheulian *megasite* has been located at Olduvai Gorge in Tanzania. Spanning an array of habitats, large amounts of lithics are suggestive of human territorially and sharing on a scope not seen before. An abundance of different stone types such as lava, quartzite cobbles and small boulders to make tools and the presence of hydrothermal features in the landscape contributed to this site becoming the most continuously occupied of the African early Pleistocene.[95]

Neural development is becoming tracked more clearly as "critical endocranial evidence" is only preserved from about 1.8 Mya. What has been found is that the all-important frontal lobes have undergone a major reorganizational shift between 1.7 and 1.5 Mya, between our one-fourth and one-third explicit nodes. Thus, changes in the brain roughly coincide with major technological developments.[96]

Given increases in intelligence, expanded home range, diversity of resource use, scavenging and/or hunting, as well as enhanced social connectivity, we are not surprised that *H. erectus* spread relatively quickly across the Old World. By the 1/4 explicit node (1.65 Mya), *H. erectus* had dispersed out of Africa into Europe and Asia. The Dmanisi, Georgia, finds, dated 1.78 Mya, show skeletal adaptations for "efficient long-travel" and the ability to bear compressive loads, such as children. [97]In Indonesia, a series of *H. erectus* finds have been dated 1.8-1.6 Mya.[98] In the Yuanmou Basin of China, a reliable documented presence for the genus *Homo* appears by 1.7 Mya. In northeast Asia, stone tools, dated 1.66 - 1.55 Mya, provide evidence for human capability to invade high northern latitudes at this early time.[99] In a related vein, hominin ranging patterns had increased to 10 km by 1.6 Mya, indicating *H. erectus* routinely accessed a variety of habitats. [100] In sum, *H. erectus* was the first human to migrate long distances, and (once settled) exploit varied ecological zones. This capacity for dispersal has been hypothesized to illustrate the evolution of a new cognitive capacity, a "natural history intelligence/intuitive biology," enabling *H. erectus* to learn rapidly and make effective decisions in new environments.[101]

At the 15% log distance from the Major Node (1.76 Mya) we have the first record of a significant development, Acheulian tools, the longest lasting technology of prehistory. Bifaced, symmetrical hand axes, about the size of a hand, have been found in Kokiselei, west of Lake Turkana, and Konso, Ethiopia.[102] These Mode II tools have been found in varied habitats and thus were the product of "generalists."[103]

By the 1/4 explicit node at 1.65 Mya, further verification of the significant technological change is found. In Kenya, Acheulian tools, co-occurring with earlier Oldowan type tools, have also been dated to 1.76 Mya and found to be in widespread use by 1.6 Mya.[104] At Peninj, Tanzania, Acheulian tools, dated from 1.7 to 1.5 Mya, show signs of woodworking activity.[105] Recent finds have confirmed that large cutting tools were used in South Africa circa 1.57 Mya, with individual samples ranging from 1.89 to 1.34 Mya.[106]

The significance of the Acheulean development should not be underestimated. These tools show the ability of humans to impose a symmetrical form on stone, called the *biface*—a large teardrop-shaped form. If there is any doubt regarding the originality of prior Oldowan tools, the making of symmetrical tools is uniquely human. Three tool types were invented and made in a standardized way: the hand axe, which comes to a point; the cleaver, which has a sharp end and a blunt end for holding; and a round discoid. The latter especially gives clearer evidence for a symmetrical sense. As a good flint knapper knows, making such tools is extremely difficult, and the end products can be characterized as works of art.[107]

Such technological advances may have served as an "ecological buffer," coinciding roughly with the onset of glacial cycles around 2.6 Mya and increased climatic fluctuations.[108] Specific tool uses have been proposed, including hand-axes that aided in animal butchering.

Some hand-axes, however, appear larger and more finely shaped than required for practical function. According to Leary and Buttermore, pleasing symmetry found in tools served as a form of mate attraction.[109] In a similar vein, Tensorer believes that the "harmonic component" expressed in bifaces illustrates *H. erectus's* capacity for symbolic behavior and artistic creativity.[110]Harrod even suggests that Acheulian bifaces—rejuvenated from cores—symbolized a "birth-giver" or the process of "being born, or perhaps reborn."[111] While such interpretations remain speculative, there is little doubt that *Homo*'s ability to impose symmetrical form onto raw material represented a significant behavioral and cognitive leap.[112] Thus, we propose that by the third hominin node, early humans could envision a precise mental construct and imprint it upon the material world. In a related vein, Buttermore and Leary place the birth of symbolic capacity around the 1.7 Mya marker.[113]

Possible examples of Palaeoart emerge at this juncture. Mary Leakey found two such objects at Olduvai Gorge. The first is a "grooved and pecked cobble," which may have been worn as a pendant, dated 1.8 Mya or at the 1/10 explicit node. The second is a hexagonal spheroid with a "dot-and-undulating-line motif, dated 1.5-16 Mya or at the 1/3 explicit node. These objects need to be examined further to verify intentional human markings.[114]

Additional evidence suggesting symbolic activity is present from the African Sterkfontein Member 5. Cut marks on a hominin skull, dated 2.0 - 1.4 Mya or circa 1.7 Mya, indicate a purposeful separation of the mandible from the cranium.[115] The special attention paid to the skull raises the possibility of rituals associated with the dead and/or cannibalism.

In the area of the 1/3 explicit node between 1.6 and 1.5 Mya, we find the most complete ancient skeleton of early humans, the Nariokotome boy. This skeleton is essentially human, with its long and slender form, adapted for the dry, hot conditions in Africa. A brain capacity, estimated at 909 cm, represents a distinct increase from *H. habilis* of the precursor period and is twice the predicted size of a primate of similar body size.[116] Thus, with the 1/3 explicit node, we have surer confirmation that humans had developed their characteristic body proportions and large brain.

Last, we note an early association, dated 1.5 Mya, between fired material, such as burned bones, with human artifacts at Koobi Fora, Kenya. While it doesn't provide explicit evidence for making fire or the controlled use of fire, it shows that fire coincided with hominin activity.

[117] This can be seen as portending regular, controlled use of fire in the following nodes.

We can now outline the evidence for this log node.

*1/3 Explicit node (1.55 Mya)Nariokotome boy 1.5-.6

Association of fire w/hominin activity 1.5

Dot & undulating line 1.5-1.6 Fire use Koobi Fora, Kenya 1.5

*1/4 Explicit node (1.65 Mya) EKMN R 1.78-1.63 Acheulian tools 1.6-1.7

* 10% Log distance (1.83 Mya) Old World dispersal 1.8 Hunting 1.8

Acheulian Hand axe 1.8 Grooved & Pecked cobble 1.8

DS (David's Site, 1.84 Ma) in Olduvai Gorge/ Base Camps

2 $^{\mathbf{20.9}}$ **❂Major Node** (1.96 Mya) *Homo erectus* 1.95-1.9 Skull 1470 1.9-1.8

*1/4 Precursor node (2.33 Mya) *Homo habilis* 2.3 Oldowan II. 2.33

*1/3 Precursor node (2.47 Mya) *A. garhi* 2.5-2.6 Early *Homo* 2.5-2.4

Cut marks on bones/scavenging stations 2.58

+ Mid-node (2.7 Mya) Early stone tools 2.6- 2.53

In summary, the third hominin node marks the appearance of the *Homo* family. The precursor period presented various stem forms with an expanded brain, along with a mix of body proportions and the first excursion out of Africa. From the Major Node and continuing into the explicit period, *Homo erectus* represents a clearer human type. *H. erectus* rapidly dispersed into Europe and Asia and created the first symmetrical tools.

We hypothesize that these varied developments went hand in hand with novel consciousness. A projecting, creative "I" exhibited the ability to fashion an impress or design upon the world. This new capability provides us with a strong intimation, yet more quantum change is on the way.

Chapter IV

4th Hominin Node; 1st Archaic Homo sapiens/ Language /978 Kya

($2^{19.9}$) 13th Node of Life

We have come to the 4th node of human evolution and to the left of our spatial chart where we look for a more hidden, quantum evolutionary change. Previous nodes that landed on the 4th node were the origin of life, the first vertebrate, the first primate, and the first human. One wonders what change could compare to these events? We hypothesize at this time the emergence of *language,* a momentous event which clearly shaped human evolution and one which is difficult to document. Our task is especially difficult as the fossil evidence is poor between 1 and .4 Mya, with few specimens in well-dated contexts.

Before proceeding further, let us pause to consider language as a unique human signature and the theories for its emergence. Although recognition can go into the hundreds of words, apes and monkeys have virtually no ability to imitate vocalizations, a key capability for language instruction. While vocal affect can be conveyed by primates, their vocal tract can't be controlled to communicate ideas. Last, while there may be a rudimentary understanding of a sequence of calls, primates lack any further perception of syntax, such as phrase structure and relationships between distant items[118]

We consider three main hypotheses for the origins of language: 1) Corballis offers a gesture *first* approach. In comparison to primal-rooted utterances from the limbic brain, he argues that hand movements are more intentionally rooted, and contain a representation of a thing and an action—the seed of language.[119] 2) Lingis offers a *sounds first* approach in which language origins are related to primate mating practices that consist of melodic, back-and-forth duets. Interestingly, language is viewed as a fall from a musical wholeness into the discreteness of meaning. Support for this theory comes from Masataka's analysis of infant communication and the finding of a "prosodic protolanguage," communication that resembles music more than language and served as foundations for both.[120] 3) Third, Bickerton favors a *meaning first* approach in which language builds up from simple meanings and composites of a few words, which are often based around social cognition, such as who did what to whom.[121] These three theories are not necessarily mutually exclusive, and could be envisioned as stages.

In any theory we should not underestimate the mother's importance. Gibson places the mother-child dyad as the seedbed for language, starting with the communication of simple needs and desires into one or two-word utterances.[122] This means that altriciality, or the increased time needed to care for helpless young, may have been the foundation for language.

Before we consider the fossil record, let us establish the ecological context. In the first node of a series, we have looked for a significant stress event and/or permissive event which helped to prod novelty. The period around the Major Node was marked by ecological stress, the most severe time between 1.2 and .8 million years ago. In Africa, conditions were not merely drier, but there were "severe arid episodes" coinciding with the development of vast savannah environments.[123] Such arid pulses and more open environments have been hypothesized to precede "key junctures" in hominin evolution. More generally, the extended period between 1 and .4 Mya was marked by an intensification of global glacial cycles, which created an environment of variability, especially in the period between glacial and inter-glacial states. In particular, Europe was gripped by harsh glacial episodes between 800 and 600 Kya, during the explicit period. In these contexts, we can imagine that enhanced communication would facilitate survival. Put another way, stress periods demanding rapid change "placed a premium on socially mediated responses, including language and innovation."[124]

Let us turn to the paleontological evidence. A recent find of the most intact female pelvis of *H. erectus*, dated 1.2 Mya, shows a pelvis wider than modern humans and a birth canal some 30% larger than formerly believed.[125] A larger pelvis and birth canal indicate infants with larger heads and thus brain expansion, which can be related to the evolution of language.

About a million years ago, Manzi describes a stage which served as a "pivotal reference" in human evolution, exhibiting a blend of primitive and recently derived morphological traits.[126] The UA 31 cranium, for example, has older *erectus* features, such as elongated shape, thick supraorbital torus. It also has a larger cranial capacity of 750-800 ccm, and *Homo sapien* traits, including thinning at the base of the skull and verticality of the parietal region. Overall, the pattern is one of increasing stature and gracility of the face and bone/muscle structure.[127] Such neotenous traits are consistent with our association between quantum change and "simplification."

The world-wide dispersal of *Homo erectus* at this time has worked to create a debate over whether the far-flung specimens constitute one or more species. Some grant the African type its own species name: *Homo ergaster,* while others label all finds from this period as *Homo erectus.* Perhaps the answer lies in the middle. An "allotaxic" model holds that the differences between populations represent distinct *geographic morphs* which retained the ability to hybridize.[128] Supporting evidence includes a 1 to .8-million-year-old African cranium, which has an endocranial capacity of 995 cc and a blend of Asian and African *erectus* traits.[129] In this view there is one overarching species of *Homo erectus* with a range of natural variation.

The presence of a debate suggests it is not yet clear whether there is more than one human species. Some years ago, Teilhard de Chardin characterized the human type as moving toward *convergence.* Campbell, Loy, and Cruz-Uribe believe that by 850 Kya fossil finds can be placed into the single species, *Homo heidelbergensis.*[130] After analyzing specimens over a million-year period, Baab found that they exhibit very similar cranial shapes, supporting a lumping of species into one type.[131] This assessment is strongly supported by evidence that variation found in these ancient hominins compares to the range found in contemporary humans.[132] Thus, even though experts use different names for *Homo* species, the metrics of variation between 1.8 and 400 K, support a single species.[133]

As morphological change is resolving into a single species, we find ourselves looking increasingly to the cultural sphere for evidence of a grade-level change. At this node we can hypothesize the emergence of proto-language. Specifically, we posit structure beyond "one call, one meaning," with at least short prompts to action that were two words long. Additionally, "little words of grammar" helped provide orientation, such as "Look, behind you!"[134]

Speech does not fossilize, and we are left considering circumstantial evidence. From the prior node, the stage for language has already been set with expanded neural capacities, cooperative hunting and meat sharing, and an abstract capability (shown in the making of symmetrical tools), as well as longer periods of maternal-infant care. At this node, *H. heidelbergensis* presents a significantly larger brain with more modern features and greater anatomical capacity for speech.

The archaeological record at this node also provides behavioral clues for the emergence of language. Between the mid-node (1.38 Mya) and the 1/4 explicit node (823 Kya), evidence for the use of fire appears. A number of African sites have burned remains which might have been resulted from intentional human activity. At Chesowanja Kenya, lumps of burnt clay, heated to the temperature of fireplaces have been found in a discrete area and are dated 1.42 Mya, along with nearby animal bones and lava tools.[135]

In the area of the Major Node the evidence for fire becomes more secure. From Wonderwerk cave secure evidence for burning in an "archaeological context" has been found at 1 Mya.[136] Early evidence for fire use by hominins also comes from Swartkrans Cave, where 270 burnt bones have been found in Member 3. Electron spin resonance has verified these bones were super-heated with temperatures usually obtained only in hearths. Originally dated from 1 to 1.5 Mya by the discoverers,[137]subsequent radiometric redating places the layer at .96 million years ago.[138] More supporting evidence comes from hominin occupation of China during an extremely cold glacial stadial at 777 Kay, near the 1/3 explicit node, for which fire would be necessary for the tropical hominin species to survive.[139] In short, a reasonable case for fire domestication can already be made in the region of the Major Node, at one million years ago.

The ability to harness and shape a dangerous form of *energy* can be seen as a parallel to shaping *matter* to make tools. With fire, humans gained the ability to transform things: to cook, to harden wood, and to change the color of stones. Cooking served to release more nutrients, supporting brain activity and expansion. Fire also provided a protected space, especially during illness and birthing. Beyond immediate survival, fire extended periods of light and warmth that intensified and focused cultural life. We can picture overlapping generations gathered around the hearth as a nursery for language.

Consistent with language emergence is the capacity for abstract representation. A series of parallel lines were found engraved on a bovid bone, dated broadly from 1.42 to 1.2 Mya, between the mid-node and the 1/3 precursor node. The marks have been described as non-utilitarian in nature and not related to butchering, so they may be symbolic.[140]

Clearer evidence for a symbolizing capacity comes from an innovation in lithics. Early evidence for a radical advance in the making of "prepared cores"—the Kombewa technique from Western Kenya has been dated at 1.1 Mya. This might be seen as a variation of the Levallois technique which produced smaller tools, and which demanded staged planning, to which we return later.[141]

In addition, finds from Member I (dated 990 Kya) from the Ologesailie formation in Africa indicate a capacity for considerable foresight. Here, adjustments in tool making took into account that tools must be carried, as they would be made thinner and re-used more often. This meant planning by hominins enabled them to better deal with scarce ecological conditions.[142]

At this time, *Homo erectus* greatly expanded their home range (the space over which they traveled in order to acquire raw materials) from 14 Km to 100 Km. Marwick believes that this expanded range necessitated symbolic communication, or a proto-language.[143] The ability to travel by sea would also supports the presence of language. In the Indonesia region, island-hopping sea voyages are documented between 840 and 700 Kya, within the explicit period. It is hard to imagine journeying over the open sea without some form of communication beyond gesture.[144]

We return to the paleontological record for early archaic humans. Although hominin fossils are remarkably sparse in Africa from 1 to .6 Mya,[145] the record is not blank. Between the Major Node and the 1/3 explicit node, a hominin from Eritrea near the Red Sea has been dated 992 - 790 Kya. "Buia man" possessed a long and oval skull with massive brow ridges, like *H. erectus,* but with a more fetal-like shape to the head in which its widest point is near the top. With a brain volume at 1000 cc, "Buia man" has convincing modern traits.[146]

More confirmation for grade level change appears between the 1/10 and 1/3 explicit nodes. At Atapuerca, Spain, researchers have found fossil hominins, dated 857 - 790 Kya. A recent dating by new techniques places the oldest hominins at TD6 even earlier, between 950 - 900 Kya.[147] Their expanded, swollen braincases evidence a "significant increase in cranial capacity" when compared to *Homo erectus,* as well as a *sapiens* type of cranialfacial architecture.[148] The brain size is 1000 cc., compared to the best preserved early African *H. erectus* skulls, which average 850 cc. The researchers describe a more sculpted, "astonishingly modern human face."[149]

Importantly, *Antecessor's* dental pattern provides evidence for an "essentially human life cycle" with an adolescent phase, one predicted at the 1000 cc threshold of brain evolution. The fully modern pattern of midfacial morphology suggests that the face of *Homo sapiens* may have appeared by a paedomorphic process in which adults retained the developmental rate of juveniles.[150] That is, future adults kept the face of the child from previous ancestors.

In terms of the human lineage, the place of the Spanish find is debated. The Atapuerca research team has named a new species *Homo antecessor*, believing it to be the last common ancestor to Neandertals and modern human lineages. Others place the finds as the oldest of the Neandertal lineage, while others classify it as *H. heidelbergensis.* These classifications are not necessarily mutually exclusive, as recent genetic and fossil finds suggest that Neandertals may have hybridized with pre-sapien types.[151]

Whatever its exact proximity to the human lineage, we sense that *Homo antecessor* expresses a new potentiality. The presence of an extended adolescence phase is relevant for language emergence, as it abetted larger brains and expanded the period for learning. It is also consistent with the association of "simplification" with quantum change. Related to this, two other traits bear mentioning . *H. antecessor* could hear an expanded range of frequencies and was right-handed. Both traits can be associated with language capacity, leading Arsuaga to hypothesize that symbolic language and reason arose by this time.[152]

One last facet of the Aupuerca finds bear mentioning. Human bones have been found which bear similar butcher marks as animals and were apparently discarded the same as animals.[153] It is possible that such cut marks represent symbolic activity, such as ritual defleshing. It may also be that cannibalism was practiced as a response to acute stress and starvation.[154] The cut marks from Atapuerca, however, continued over long periods of time, even when plenty of food was available.[155] The evidence for cannibalistic practices among early humans is sporadic yet widespread and questions related to its origin and practice remain unresolved.

We return to evidence for cranial expansion, which is widespread at this time. The 900 to 800 thousand year old Ceprano skull from Italy shows "marked widening of the frontal lobes," when compared to earlier erectus skulls, suggesting it could be an "ancestral stock" to *H. Heidelbergensis*.[156] Dated 780 Kya, the Yunxian Skull from China has been reconstructed by Vialet to show an endocranial capacity of 1050 cc.[157]

Returning to Africa, morphological evidence is supportive for the emergence of language. By 759 Kya near the 1/3 explicit node, the shape of *Archaic sapien* skulls has the capacity to make the full range of sounds needed for language. Additionally, the pharynx size is sufficient to make vowel sounds, and the mouth is small enough for the tongue to make consonants.[158] Thus, the biological basis for hearing and speech have come into place.

The 1/3 explicit node offers yet more confirmation for a leap in cognitive capacity and language emergence. The remarkable site of Gesher Benet Ya'aqov, dated 780 Kya, show multiple proofs for a new level of complexity. There is evidence for the dismemberment of an elephant, a worked basalt core, hand axes, and what appears to be worked slabs of wood—where the death site has become an "integral part" of the living floor.[159]The remains of the elephant may be the first proof for big game hunting. The hunt of large game required a sophistication gained from learned behaviors acquired over generations.[160] In terms of harvesting plants, seven species of nuts were cultivated, and tools were found which exploited them, such as pitted hammers and anvils.[161] Varied food production and an intricate domestic space intimate a new level of attainment.

By this time, the "most compelling evidence" for the early use of fire can be found at Gesher, in the form of multiple clusters of burnt flint and wood."[162] "Latent hearths" are documented by spatial analysis, revealing the routine use of fire. Heated lithics also reveal an association between percussive activities and fire.[163] Thermoluminescence analysis is confirming of hearths, showing a pattern of increasing intensity of fire the closer the lithic is to the fire's center.[164] Last, fire use is *continual* throughout Gesher's long occupational sequence in which 13 archaeological levels approximate 100,000 years. This means that hominins along the damp lake shore were able to start fires at will. [165]

Perhaps the most significant technological innovation from Gesher is a change often ascribed to the next node. Several huge basalt cores have very large flake scars, indicating the presence of the *Levallois* technique (Mode II technology) by 780 Kya.[166] Sharon cited to me that the Gesher site "undoubtedly has Levallois cores and flakes,"[167] and has since published an exhaustive compendium of the site which documents this.[168] The Gesher site is consistent with our hypothesis for language emergence at this node, as it demonstrates "adroit technical skills and in-depth planning abilities."[169] Gonen Sharon put it this way, "Although it can't be proven scientifically, personally, I am sure they had language and I see no other way by which such sophisticated tools and life ways would happen."[170]

Similarities to older African traditions with "approximate Levallois patterns" show that this tradition was likely the result of an earlier wave out of Africa. The earliest evidence for Levallois technology in Africa appears to be a small prepared core, also dated 780 Kya, from Wonderwerk, South Africa.[171]

The Levallois technique can be described as the third major advance in making tools. With "staged toolmaking," hominins demonstrate they can think ahead and develop an intermediate product.[172] In this technique, a core is initially prepared, after which blades are struck off by a single hit. The end result is a V-shape form with sharp edges on either side. This meant a ready way was devised that could impose a congruent and three-dimensional symmetry on stone tools.[173] Last, percussion marks on wood document the use of a soft hammer, making for thinner and more mobile hand axes.

There is debate over whether the lithics previously cited are strictly speaking the Levallois technique.[174] The reality is that close examination of the modes of making tools reveals a variety of intricate prepared core methods, sometimes at one site. The definition of the Levallois mode is one example and requires a highly technical definition. Some of these lithic types, in carefully analyzed sites from the next node, have been hypothesized to be "proto-Levallois."[175] What we can conclude is by the 4th node, there is a significant advance in the complexity of techniques, in which "prepared cores" were used to make tools.[176]

We ask at is point, what can the making and use of such tools tell us about "symbolizing humanity?" We can articulate the steps of making such tools in three steps. The first relates to the base or foundation, envisioning a potentiality in a large stone at the point of gathering the raw material. The second is taking the complex series of steps to actualize the design. This relates to the third, a vision or image in the mind throughout the process, which guides your task and propels you to complete the task. The end result is a beautiful product, which then enters and acts upon the world.

I wonder if these steps can be related to the Indigenous concept of three worlds. The base or foundation concept can be seen as related to the "underworld," a world which offers the raw lithic material and a sense of its potentiality. Second, the series of steps taken, which first involves shaping the core before a second shape is derived, can be related to "This World" activity, and one which makes use of the hands. Last and third, the "overworld," an overarching dimension, can be related to the mental vision, held in the hominins mind, of a specific end product. This vision is related to beauty, which is expressed in the end product. Last we note that the effective use of the tool then goes on to interface with an alter This World.

We offer this analysis as not as a proof to the origin of Three World concepts, found in many Indigenous people across the world. Rather, we offer the suggestion that it is good to think about how practical activities may resonate with human conceptualizing of the world.

Even though we focused here on the capacity to make lithics to illustrate a leap in symbolizing humanity at this node, we should keep in mind that tool making is but one aspect of a larger, more complex reality. Returning to Goren's work once again, we cite at some length the conclusion of his study.

> ....the continuous culture documented at GBY not only required a fully developed language, but was entrenched in

> a distinctive, consistent, continuous, and homogeneous tradition, transferred through many generations and expressing an extremely diverse and impressive culture. ...these cultural expressions were not, as is frequently claimed, features of a typically stagnant Acheulian culture, but on the contrary a highly variable culture with multiple material expressions, some of which can be interpreted and understood as parallels to "modern behavior."[177]

We now review the evidence in chart form:

+ Mid-node (691 Kya)

*1/3 Explicit node (777Kya) Gesher /Levallois tools 780 Sea travel 800

Hominins in China, during extremely cold glacial stadial 777

*1/4 Explicit node (823Kya) *H. antecessor* 780-857 Pigments 8-900

*10% explicit node (913) Earliest TD6 hominins 900-950

2 $^{19.9}$ ❂ Major Node (978 Kya) Swartkrans cave fire/superheated bones 960 Kay

> "Buia man" Stratigraphy 992 Paleomagnetic >1.072 UA 31 1.0

*1/4 Precursor node (1.16 Kya) Early "prepared cores" 1.1 Trade networks expand 1.0

*1/3 Precursor node (1.23 Kya) Tally marks 1.4-1.2 Arid spell 1.2-.8

+ Mid-node (1.38 Kya) Early evidence for fire 1.42

In summary, the 1st archaic *Homo sapien* node presents us morphological and behavioral novelty, which leads us to hypothesize the emergence of a proto-language. Evidence includes marked brain expansion, an adolescent phase, expanded networks of trade, a sea-faring capability, and extensive use of red ochre. If the precursor period presented the first evidence for the use of fire, the explicit period brought certainty for the continual use of fire. In terms of tool-making, the prepared core technique, clearly seen as Gesher, represents a marked conceptual advance. Thus, on several scores, we can make a reasonable case for the emergence of a proto-language by this node.

We look forward, however, to the next node, where we expect yet clearer evidence for a speaking, symbolic humanity.

Chapter V

5th Human Node; 2nd Archaic sapiens/ Language / 489 Kya

($2^{18.9}$) 14th Node of Life

With the second node in a series, we look for a heightened social fusion and spread into new geographic zones. As with each node, we expect marked enhancements in the neural sphere and/or increases in behavioral complexity. In particular, we look for clearer signs of symbolic activity and a capacity for language.

Let us first turn to fossils of archaic *sapiens*, which present evidence for further gracilization: reduction of brow ridges, jaw size, and chewing teeth, and a general lessening of the frontal projection of the face. With finds centered in Europe, *Homo heidelbergensis* is accepted as a species with its type specimen dated to 609 Kya ±40 thousand years, between the 1/3 and ¼ precursor nodes. Out of a plethora of species being named during this time, *H. heidelbergensis* has been proposed as a good candidate for the Last Common Ancestor of Neanderthals and modern humans.[178]

Although much of the above has focused on *H. Heidelbergensis* as a significant species, we add this caveat. The naming of new species at this node may exaggerate subtle differences found in the converging human type. Congruent with this, most specialists refer to the many labeled species at this time as "archaic homo sapiens," confining *H. Heidelbergensis* to the European finds.[179]

At this node brain expansion of the archaic sapiens species make a disjunct leap. Compared to the late *Homo erectus*, who averaged 950 cc, brain size has increased to approximately 1200 cc. The widest part of the brain is now in the sides, the parietal regions, rather than the cranial base.[180] After an examination of dozens of named *H. heidelbergensis's* skulls, cranial increase—relative to body mass—represents a "significant" increase over *H. erectus*. The "encephalization quotient" (EQ) is now 4.83 compared to 3.61, a 25% increase, which is characterized as a disjunct leap.[181] Barham and Mitchell have placed the speciation event for *H. heidelbergensis* at 600 Kya, coinciding with a significant increase in absolute and relative brain size.[182]

Speech may not fossilize, but there are specific anatomical indicators. In the region of the 1/4 precursor node, by 600 Kya, sufficient breathing control had arisen to produce speech.[183] With the European *H. heidelbergensis*, the bony palate, the sole anatomical structure used for producing language and which is often fossilized, takes on modern form.[184] Another indicator for language is neocortex's ratios, which show an exponential increase in the group size for *H. Heidelbergensis* (to 150 individuals), making it likely they needed language to maintain social bonds.[185] In summary, the ability to produce and hear language, along with larger group size, strongly suggests the presence of language by the 5th node.

At this node, more direct biological evidence for language mounts. The Atapuerca craniums exhibit a modern human capacity for hearing, with heightened sensitivity between 1 and 3.5 kHz and above, very near the human range.[186] Two hyoid bones, throat bones involved in speech, have taken on modern form in morphology and dimension. [187] A recent reconstruction of *H. heidelbergensis's* middle and inner ear indicates sensitivity to frequencies between 3 and 5 kHz, the range in which language occurs.[188] A temporal bone of *H. heidelbergensis* dated 500-300 Kya, during the explicit period, verifies a hearing capacity similar to *Homo sapiens.* Last, a large orifice for a facial nerve suggests "unusually complex facial musculature" associated with facial expressions.[189]

We go o Africa for evidence of brain expansion and culture. From the Middle Awash region of Ethiopia, the Bodo Skulls were originally dated c. 630 Kya, near the 1/3 precursor node. A recent revaluation, however, places the event c. 523 Kya, near the 10% precursor node and quite proximal to the Major Node.[190]These early fossil humans, with a mix of *Homo erectus* and *Homo sapien* traits, have a cranial capacity of 1200-1300 cc. Some class this find as *H. heidelbergensis*, others grant it their independent status as their own species, but we note again they may better described as archaic *Homo sapiens.* Importantly, their skulls exhibit signs of cultural alterations. Cutmarks around the eye sockets, forehead, cheekbones, and elsewhere indicate defleshing in a non-utilitarian fashion, possibly indicating a mortuary ritual.[191] Hayden believes this practice reflects the belief of a "spirit essence" located in the head, and a sense of a soul or afterlife.[192]

From a bovine shin bone from Kozarnika cave in Bulgaria, regular, repeated marks have been identified. This bone, found in Layer 12, is generally dated as greater than 500 Kya, and less than 600 Kya, [193] This provides some of the earliest evidence for symbolic activity, although the meaning of four sets of four marks are unknown. We note, however, that a report of this find offers us a useful definition of a symbol, as "a visible and tangible reality suggesting other invisible realities."[194]

In the vicinity of the Major Node, clear evidence emerges for hominin use of red ochre.[195] Dated circa 500 Kya, pieces of red ochre, showing signs of use, have been found at Kathu Pan, South Africa.[196] The nearby Wonderwerk Cave also provides suggestive evidence. At every level of the cave, the substantial occupation sequence has produced an abundance of ochre fragments along with

Acheulian bifaces and exotic quartz crystals. The purposive mining of red pigments and the making of crayons leads Bednarik to speak of "conscious mark production" and color preference.[197]

Further evidence comes from slabs of ironstone that can be classed as ochre, and which show incised lines. These objects were found in the back of the cave from Excavation 6 and classed with the Fauresmith Industry, generally dated to ~500 Kya.[198]

The finding of such a treasure trove in the back of a 100-meter cave, the length of a football field, is intriguing. Although a circle of light is visible from the back, the environment is generally one of darkness, silence, and resonant sound.[199] We can ask, why were exotic manuports such as crystals and pebbles brought such as distance away? It recalls to my mind a deep shaft, with a ring to suspend ropes from, which I was shown by elders in Guatemala, as a place of youth initiation. Youths in this Pleistocene cave may have been taken into the depths of darkness, carrying exotic objects, and leaving them behind as a token of their journey.

The symbolic use of red ochre is widely documented among Indigenous cultures and has a long prehistory. Its potential associations include life, vigor, blood, fertility,[200] wounds, and a passage from death to life.[201] The instance, the Maya used red pigment to cover bones because red was associated with the east, where the sun is reborn every morning. Although red ochre was also used for practical purposes such as tanning hides, its wholesale use and the making of crayons implies symbolic use. Along such lines, Hayden believes that early red ochre use most likely held "supernatural connotations."[202]

It's worth pausing even more over the use of red ochre, a substance that penetrates into the deep time of humanity and enjoys worldwide use. From its most early presence, we can infer that ochre was put on something, transforming its appearance. This reminds me of when my son at a young age discovering purple pokeberry juice, smeared it on his hands and stared at amazement at his hands. A dying substance can paint, anoint, sacralize, and symbolize an object. From hands the substance can put on tools, decorative beads, the walls of habitations, or the body. This action creates another view of reality which opens a door for the imagination. Here, near the center of our logarithmic ten nodes, we have come upon a major symbolizing substance of humanity.

This highlighting of ochre should not detract from the beauty of black, the use dark flints, the luster of obsidian, as well as the place of white as found in shells. The presence of brilliance, which includes sparkling forms of ochre, is an aesthetic quality that invites a transforming vision of the world. [203]

The capacity for symbol making gives us pause to consider further this general trait of human nature. Symbol use can be said to be a defining human trait critical to our ongoing evolution. Hogenson puts it this way: symbols are devices with which we *bootstrap* up into a human world. A symbolic world means that our world can become an "infinite world," one that is paired with a self that is "constantly emerging," passing through "potential states" and phase transitions. [204] Congruent with this is Hensilwood's assessment, that the ability to encode and store data outside of the human mind marks the most important turning point in modern human behavior.[205] Humans, it could be said, have used symbols to accelerate their own evolution.

Technology, represented by lithics, shows significant gains. Around the Major Node, evidence appears for the first hafted tools, in which a formed stone tip is attached to a wood handle or stick, such as a spear. From Kathu Pan South Africa, retouched points dated to ~500 Kya, have been identified as forming spear heads.[206]

Dated between 510 to 284 Kya (between the Major Node and the next precursor node) we can place the Kapthurin Formation in Kenya. This early Levallois site possesses biface preforms, a diverse array of Levallois and blade lithics, and extensive processing and use of red ochre. Deino and McBrearty believe that such evidence points to "profound changes in hominid behavior."[207] Significantly, the first evidence for blade production, generally dated at 500 Kya,[208] has recently been more precisely dated between 465 and 395 Kya, between 5% and 1/3 log distance after the Major Node.[209] Blades, which are long, thin and sharp, represent a leap in complexity in terms of their making, requiring a much greater number of blows to produce the final object.[210]

From Gesher Benot Ya'aqov, yet another significant find has been dated to within 5% log distance of the Major Node. The earliest evidence for the structural use of wood is provided by two planks with interlocking notches, dated to 476 Kya.[211] The ability to plan and construct something with two parts, as exhibited by the earlier mentioned hafting, or more represents another milestone in humanity's conceptualization/actualization.

Between the 1/10 and one-fourth explicit nodes, we can place a trove of hominin fossils from Sima de los Huescos of Atapuerca, Spain. The finds exhibit a mix of ancestral and derived features and are classed as either a late form of *H. heidelbergensis* or early Neandertals. As suggested earlier, many traits of these hominins are shared with the cold-weathered Neandertals who dominated Europe in later nodes. These include very robust bodies, large cranial capacities, and sexual dimorphism in the contemporary human range. DNA analysis, which has been able to accurately sequence the Neanderthal genome, has recently verified early Neanderthal ancestry.[212] Neanderthals, whose first find was from Neander Valley in Germany, have received underserved bad press, as their accomplishments match those of their African counterparts.[213]

Recently dated to 427 Kya,[214] numerous fossils were found in a context which strongly suggests ritual activity.[215] A set aside burial pit was located in a difficult to access cave system, some 600 meters into a cave, down a 13-meter shaft, and in a select alcove. It contained some 32 human remains, mostly of adolescents and young adults who died about the same time. The ages of the individuals, and the fact that several generations were regularly placed away from an occupation area, points toward a culturally based selection, or a burial.[216] Additionally, the placement of a beautiful, light red-brown Acheulian Hand axe (made of good quality, veined quartz) along with the human skeletons in the pit appears to confirm ritual intent.[217] Cultural mediated burials grant a sense of connectivity between the living and the dead, consistent with our expectation for a novel social fusion at this node. Congruent with this, Barbara King has hypothesized the sense of *belongingness* moved beyond the family and kin group to include those who had died—forming the seedbed of religion.[218] In sum, the earliest evidence for burial, a momentous indicator of humans being conscious, has occurred by the 5th human node.

With the second node in a series, we look for expansion into novel and more stressful geographic zones. About 500 Kya and near the Major Node, humans colonized colder realms in Europe, necessitating the regular use of fire. In Britain, a hearth, dated from 465 to 380 Kya, offers definitive proof for controlled use of fire: burnt bones and charcoal in a ring of stones.[219] By 400 Kya, a number of sites in mainland Europe provide evidence for the regular use of fire.[220] The more routine and systematic use of fire, as we previously mentioned, would accelerate cultural activity, as groups gathered for long hours to shelter from the winter cold. Lumley believes that structured fireplaces, dating 400 Kya, served as an "extraordinary motor of hominization," strengthening group spirit and allowing for the birth of the first myths.[221]

By the explicit period, evidence for sophisticated big game hunting emerges. A camping site from Schoningen, Germany, dated 400 Kya, has spears, wooden shafts for mounting tools, and repeated cut marks on large bison bones, indicating meat was sliced for drying.[222] Three javelin-like spears for throwing were found in association with stone tools and thousand of butchered remains of horses, providing convincing evidence for *big game hunting.* The invention of the spear allowed humans to kill at a distance and shifted the odds to their favor in encounters with large animals. Along these lines, Stiner posits a "threshold shift" in animal exploitation around 500 Kya involving large game hunting, increased group cooperation with the use of spears, and the invention of fire roasting.[223]

Direct evidence for symbolic activity appears during the explicit period. At approximately 400 Kya (between the 1/4 and 1/3 explicit nodes), a large grindstone from South Africa was incised with checker-board, crisscross lines.[224] A signature site from Bilzingsleben, dated 420-350 Kya, provides us with four bones bearing imprints of lines in a form of rhythmic sequence. Davis finds the evidence sufficient to characterize the start of a style or tradition.[225] Bednarik, for his part, describes the metatarsal bone as incised with a "very complex arrangement," involving a "more advanced concept-mediated marking. [226]

On a riverbank in Tan Morocco, Africa, a possible modified sculptural-like stone, was found near stone tools, dated 500-300 Kya. The red quartzite stone has been described as a figurine of indeterminate gender and, more conservatively, as a "modified object" of no utilitarian value. Bednarik has identified eight man-made grooves that accentuate its already anthropomorphic appearance.[227] Bahn cautions us that this interpretation remains dubious.[228] A coating of bright red pigment, the earliest evidence for applied coloring material, however, would appear to confirm cultural intent.

We now review the archeological evidence in outline form.

*/1/3 Explicit node (388 Kya) Tan Morocco "figurine" 500-300

*1/4 Explicit node (411Kya) Symbolic markings Bilzingsleben, 420-350 Spears 400

10% 456 Blade Lithics Kapthurin Formation 465-395

5% 472 Earliest use of structural wood/logs Zambia 476 plus/minus 23

2 [18.9] ❂Major Node (489 Kya) Cold zones 500 Hearths 465-380

1st hafted tools c 500 Marked Bovine tibia Kozarnika Cave, (Layer 12) >500 and<600)

*10% 524 Bodo skull with ritual cut marks 523±42

*1/4 Precursor node (582 Kya)

Atapuerca/burial 600 H. Bodeoenis c 600kya

*1/3 Precursor node (616 Kya) Homo Heidelbergensis type specimen 609 ±40 Kya

+ Mid-node (691 Kya) Hearth in Israel (790-690)

In sum, the evidence for symbolizing humanity has become more assured by the 5th node. Evidence includes markings on human skulls, abstract markings on animal bones, the use of red ochre, and a possible sculptural form. Varied anatomical indicators for a capacity to hear and speak become present. Last, evidence consistent with language and a heightened cultural life includes big game hunting and the routine use of fire. We are now ready for the 6th node, which is third in a series of four, in which we look for yet further elaboration and a flowering of speaking, symbolic humanity.

Chapter VI

6th Human Node; 3rd Archaic sapiens / Language/ 245 Kya

($2^{17.9}$) 15th Node of Life

With the 3rd node in a series, we expect a significant elaboration in the cultural sphere, more social fusion, and signs of resilience as demonstrated by technological gains and/or geographic spread. We will also look for more sure confirmation of the presence of symbolic capacity and language. In the period surrounding the Major Node from 300 to 100 Kya, genetic evidence indicates at least one strong population bottleneck in Africa, in which human population was reduced between 50 to 90%.[229] Such an event would have been a strong impetus for increased dependence on social systems, technological gains, and language.

We first consider evidence from the morphological domain around 300 Kya, close to the 1/3 precursor node. The famous skull from Broken Hill in Zambia has recently been dated to 299 Kya, plus or minus 25 K.[230] Originally called "Rhodesian Man," a remarkably complete cranium indicates a brain size of 1280 cc, a brain size we saw in the previous node. Although traits resemble those of *Homo ergaster*, the specimen's dental and postcranial morphology is so modern that it has been described as belonging to a subspecies of *Homo sapiens*.[231]

From Jebel Irhoud, Morocco, two fossil cranium, 1300 and 1400 cc sized, reveal enough traits to be characterized as the earliest known member of *Homo sapiens*[232] or part of the lineage leading to them.[233] Recent finds of fossils and stone tools at the site have securely dated it from 315 to 286 Kya, in the area from the 1/3 to the ¼ precursor nodes.[234] In line with the above evidence, we can say that brain size is generally falling in the range of modern humans by 300 Kya.[235] However, and importantly, the shape of the modern human brain, with all the reorganization that is implied, has not yet been obtained.[236]

We approach the 10% precursor node, looking for more verification of grade level change. From Lake Turkana, Africa, fossil cranium (NM-ER 3884), dated 270 Kya, are a "near modern" find, with an estimate of cranial capacity at 1400 cc. [237] Just within 10% of the Major Node at 26 Kya, from an open site in South Africa, the Florisbad skull has been designated as *Homo helmei.* [238] This date is not definite, however, as the contemporaneity between a dated tooth and a *H. helmei* clavaria is uncertain.[239] This species possessed a large face and has been described as a precursor of modern *Homo sapiens*, intermediate between primitive archaic *sapiens* and the latter more derived *H. sapien.* Because of the fossil incompleteness, its cranial capacity is imprecise spanning between 1280 and 1450 cc. In varied ways this find has a precursor quality, including the skull displaying,

> ....a mosaic of derived and plesiomorphic features, which makes this fossil compatible with distinct phylogenetic scenarios.[240]

Not all happens in Africa. Also around 260 Kya, a fossil from Jinniushans, China presents archaic species. At 5 ½ feet and 175 pounds, an archaic female possessed a cranial capacity of 1,300 cc with an EQ of 4.5.[241] The reference to China, mentioned several times prior, deserves further note. Rather than an evolutionary backward as historically portrayed, we see an evolutionary dynamism continuing at work. Around 300 Kya in Asia, as evidenced by the Jinniushan fossil find, "a critical turning point" in morphology is occurring during a period of climatic instability. Rather than stasis or one suite of traits, the finds at this time are characterized by "increased morphological variability."[242]

The last observation calls to mind what has been characterized as astonishing finds. In a South African cave, hominid fossils have been dated to 335- to 23, spanning the precursor period. Deep in a cave, 90 meters in a dark is a chamber with the remains of 15 individuals, who present a species unlike any other. They have a mix of primarily Australopithcine traits, but include *Homo* and *Paranthropus*, as well as their own unique traits. Their brain shape is like *Homo erectus,* but its much smaller size goes back to *Homo habilis* dimensions, which is about one third of contemporary humans.[243] In our charting we are not surprised to find a precursor period filled with experimental and various results.

In sum, the precursor period presents a dynamic period in which finds are often seen as intermediate to *Homo sapiens*. As we approach the Major Node, cranial evidence of increasing brain size is suggestive of a grade level change. Brain size is now falling within the range of contemporary humans, whose average brain size is 1400 cubic centimeters. From here, morphological change will be focused upon the shape and complexification within the brain.

From bodily changes, which are hard to read, we turn to what humans produced, and how they modified objects. By this time, the Levallois technique spread to Europe and exhibits considerable variety. In France, Levallois tools and specific work activity sites are found in caves dated 300 Kya.[244] In Africa about 250 thousand years ago and at the Major Node, the Koimilot assemblages of the Kapthurin formation show considerable variability in the production of Levallois flakes of different types.[245] In one estimate, the number of identifiable tools in this era has quadrupled to about forty types.[246]

The use of color has become more pronounced, as ochre fragments become associated with tool artifacts and animal bones from a number of sites. At Twin Rivers, Zambia (400 - 266 Kya), systematic use of ochre occurred in five different colors, including one which sparkles.[247] These pigments are hypothesized to have been employed during birthing rituals to deal with the stresses of birthing brought on by increased head size of children around 300 Kya.[248]

The Berekhat Ram "figurine" has been broadly dated between 470 and 233 Kya, with the excavators favoring 280 - 250 Kya. Careful microscopic analysis of this bright red tuffic stone object, about 1 ½ inches in height, reveals alterations that enhance a figurine appearance.[249]The Israel museum of Jerusalem states that the same flint tool was used to shape the "head, arms, and ample bosom."[250] Its shape is also suggestive of pregnancy,[251] which could relate to symbolic acts and representations seen as generating life.[252] The application of red ochre to the modified stone is confirming of a symbolic intent. Such early works have been classed as "nature-anthropogenic objects," in which suggestive natural shapes are minimally altered by hominins to create a symbolic object.[253]

Last, and in the area of the 1/3 precursor node, we come to a find not securely dated, In an Austrian cave, archaeologists have discovered a perforated wolf incisor, possibly dated to 300 Kya.[254]Body adornment suggests an emerging sense of personal identity and social demarcation.

From the 1/4 precursor node to the Major Node, more complex composite tool production emerges, in which stone, handle, and adhesives were combined. Evidence for deliberate backing of small blades or hafted tools comes from Kalambo Falls, Zambia (265 Kya), and the Kapthurin formation (235 Kya). The making of tools from multiple components, as intimated earlier, requires a high level of skill, imagination and intentionality.[255] Whether for butchering or digging, extensions for stone tools brought a logarithmic increase in force. We can also imagine such tools becoming objects for social display.

An indicator of enhanced social bonding is evidenced by the regular use of *hearths,* dated near the Major Node (250 Kya). Scholars have hypothesized that tubers are now being cooked, releasing nutrients and calories from hard-to-digest carbohydrates.[256]

The evidence for big game hunting becomes indisputable by 250 Kya. On the Channel Island in Jersey, hunters drove wooly rhinoceros and mammoth off cliffs, then butchered them and ate their meat. Any advance in food production is of evolutionary significance, considering its impact on social life and survival in scarce times.

Near the Major Node and dated 300-200 Kya, early claims for *rock art* come from cup marks and a meandering line from a deep cave in India.[257] Bednarik suspects they may even be older, as the finds lay below Acheulian and Oldowan-like chopper tools.[258] In addition, a cupule and meandering line on a quartzite boulder, as well as a natural standing stone with 7 deep cupules, were found in a center of a chamber and have been dated to this early time.[259] Cupules are the earliest designs found in rock art and are found on all continents except Antarctica. Often groups of cupules are found, pounded into hard stone. Possible meanings range from fertility to solar associations. The making of cupules may have possessed significance in itself, as in Indigenous Australia, where hammering was believed to be an act of impregnation.[260]

In the area of the 1/3 explicit node (194 Kya), more gains are found in the technological sphere, as well as evidence for an aesthetic sense. The lithic production of hafted core axes in the Sudan indicates that there was economic specialization and possibly division of labor by

200 Kya.[261] From El Greifa, Libya, also dated to 200 Kya, ostrich eggshell disc beads are being perforated, providing confirmation for personal adornment.[262] Finds such as these serve as a sure indication of a grade level change, which Bednarik describes as "a level of hominin self-awareness that essentially expresses full cognitive modernity."[263]

Let us review the evidence in chart form:

*1/3 Explicit node (194 Kya) Standardized lithics 200 Ostrich disc beads 200

*1/4 Explicit node: (206 Kya) Rock art 300-200

10% (228) Omo fossils/Earliest *sapien* fossils E. Africa 233 ± 22 Kyr.

$2^{17.9}$ ❂ Major Node (245 Kya) Striations 250-300 Berekhat Ram "Figurine" 280-250

Homo naledi (335-226)

*10% log distance (262) Florisbad Skull (259) Hafted tools 265-235

NM-ER 3884 Lake Turkana, Kenya 270

*1/4 Precursor node (291 Kya) Possible body adornment 300

Middle Stone Age begins (c 300K)

*1/3 Precursor node (308 Kya) Levallois 300 Turkana find 300-270

Broken Hill Skull 299 +/- 25 Kya

Perforated wolf incisor/Adornment 300?

+ Mid-node New estimates for time of *H. Sapiens* divergence (350-200)

In sum, the 3rd *Archaic sapien* node grants several firsts in terms of cultural achievements: composite tools, a 3-dimensional "art" object, polychrome use of colors, rock art, and personal adornment. We sense cultural evolution has reached another threshold, in which associations become more manifold and complex. We are now ready for the last node of this series and another quantum change.

Chapter VII

7th Human Node; 1st Homo sapiens / Complex syntax 122 Kya

($2^{16.9}$) 16th Node of Life

At every 4th node we look for a more quantum, and relatively invisible, evolutionary change. At this time we can mark the appearance of *Homo sapiens*—a finalization of human form with a more gracile look. Given the prior substantial changes in the cultural and technological dimensions, we look for even more significant innovation. Here, we hypothesize a second stage in the evolution of language: the birth of *complex syntax* and the capacity to generate unlimited meaning.

The technical words for a complex syntax in language can be complex themselves: phrasing, embeddedness, and long-distance meaning. The meaning, however, can be simply shown. There's considerable difference between saying: "Look, a leopard is ahead of us!" to "What if a leopard appears when we go out on the hunt?" Along such lines, researchers have hypothesized that an *Enhanced Working Memory,* critical for language, arose between 100 and 50 Kya. In this view, *introspection*, represented by "what if" statements, the future tense, and more complex syntax jump started a cultural explosion by *Homo sapiens*.[264]

Other important related developments likely preceded or accompanied the emergence of a syntactical language. Calvin believes that language reflects the capacity for "structured thought," including the ability to construct narratives, make music, perceive chains of logic, find hidden patterns, and do in-depth planning.[265] Most likely, a range of associated capacities arose in tandem with this fuller emergence of language.

Let's now consider fossil finds and the arrival of the Anatomically Modern Human (AMH), better known as *Homo sapiens*. The face, at this time, can readily be recognized as close to our own, with diminished brows and a more globular brain tucked under the face, as well as facial reduction and the appearance of a chin. In tandem with this, from 115 to 110 Kya, in the region of the Major Node, researchers have identified a stage II of "brain globularity."[266] The shape of the brain has become more critical than the size, indicating a modularity and complexity which is approximating the 50 electrical fields of the contemporary human brain. Last, we note with *H. sapiens* the degree of cerebral asymmetry, a unique human trait tied to language, has become strongly developed.

While an African origin for *Homo sapiens* is likely, the timing is uncertain. Molecular estimates for their first divergence range widely, from as far back as the mid-node (172 Kya), to even earlier. For example, one estimate places our "most recent common ancestor" at 171.5 Kya, plus or minus 50 thousand years.[267]

Whether *Homo sapiens* came out of Africa and replaced older types, or migrated out of Africa and integrated with archaic lineages, is also debated. In favor of the latter theory, Relethford hypothesizes that moderns emigrated in waves and hybridized with local archaic populations making for successive "morphospecies," *Homo erectus*, *Homo heidelbergensis*, and *Homo sapiens*.[268] Supporting evidence includes an inconsistent mosaic of skeletal remains between 150 and 50 Kya, meaning that the "dominant pattern" was an intramixture of expanding modern humans and regional archaic populations.[269] In short, hybridization remains a viable scenario for the origins of *Homo sapiens*.

The dating for the first fossil evidence for *Anatomically Modern Human* is still being established. "Modern" human skulls from Omo-Kibish were initially put at 130 Kya by Uranium-Thorium dating. Then Argon/Argon dating constrained finds to between 195 and 104 Kya, with the preferred date close to 195 Kya, going back to the mid-node.[270] Yet another study places this find to 233 Kya and thus the prior Major Node.[271]Although some skulls clearly look modern, as a sample, they show a mix of modern and archaic traits. Importantly, they had not obtained the shape or globularity of the contemporary *Homo sapiens.[272]* Thus, in the region of the mid-node or even earlier, we find early transitional forms to *Homo sapiens*. Although the name may be given, we would not yet posit sufficient evidence for the grade level change.

A little prior to the 1/3 precursor node (154 Kya), intriguing evidence suggests that the human developmental cycle had changed. Dated 160 Kya, a juvenile mandible from Jebel Irhoud, Morrocco indicates a pattern of tooth eruption comparable to modern *Homo sapiens*.[273] A period of childhood with a modern length is congruent with our expectation for the presence of full language capability. In the region of the 1/3 precursor node, *Homo sapiens*-like skulls have been found in Herto, Ethiopia. Dated 161-150 Kya, this coincided with a cold, dry interval and a significant population decline.[274]The discoverers describe "intermediate" forms between the Bodo skulls of two nodes earlier and later Lasies and Qafzeh skulls. Exemplifying a population that is "on the verge of anatomical modernity" with the presence of some primitive traits, researchers have given them a palaeosubspecies status, *Homo sapiens idaltu*.[275] Cranial expansion is clear with the Herto skulls, as their proportions show a modern "high cranial vault" and large frontal and parietal sagittal dimensions. The overall endocranial volume of the skull has reached 1450 cc, the high end of the human spectrum.

The Hereto craniums bear signs of cultural activity. Three of the skulls were inscribed with intentional marks, or "postmortem cultural modification" by sharp obsidian flake edges. The marks appear to result from defleshing, which was non-utilitarian in nature.[276] Additionally, the sides of a child's skull display polish from repeated handling. The non-practical and purposive alteration of skulls invites us to infer ritual activity related to death.[277] We can hypothesize that humans are becoming conscious of death, for which rituals helped alleviate the attendant anxiety. Furthermore, in Indigenous cultures, rituals are "virtually always" accompanied by petition or prayer.[278] Thus, mediating the stresses of life passages may have served as an early stimulus for language.

Evidence supportive for fuller language emergence comes from an analysis of human networks. Marwick hypothesizes that "language abilities become visible" by 130 Kya, near the Major Node, a point he deduces from the exchange of raw materials over a range of 340 to 140 kilometers.[279]Whether for gift-giving or bartering of materials, language would facilitate more complex human interactions.

We expect further signs of a symbolizing humanity, if language is taking root. At the 1/4 precursor node, dozens of shells have been purposively perforated, showing signs of polish and striations. These remarkable finds, dated 142 Kya, further confirm human adornment in the form of necklaces. Such objects show all three of our indicators for grade level change. The first is consciousness, already referenced, due to the demonstrable use of symbolic objects which convey a sense of identity. The second illustrates mobility and a sense of connection to landscape, as these finds came from the ancient coastline, 30 to 50 km away, and then were brought "several meters" into a deep cave. The third is that the proliferation of such objects demonstrates "expanding social networks and a complexity of social interactions."[280]

Further indicators of symbolic activity are intentional, patterned markings. At the site of Oldisleben, Germany, two engraved bones dated 135-80 Kya, possibly at 120 Kya, bear two rows of sub-parallel grooves. A third bone from the same site has an arrow or dart-like form which prefigures upper paleolithic abstract signs. Regarding the latter design, Bednarik believes we are presented with the "oldest two-dimensional iconography."[281]

By the 1/4 and 1/3 explicit nodes at 103 and 97 Kya, we expect clear confirmation for grade level change. At this time, samples from Skhul and Qafzeh, Israel (dated 101-95 Kya) have a pattern of overall modernity with a minority of retained archaic features.[282]The average cranial capacity is 1,550 cc, easily within the range of modern humans. Significantly, these two Mousterian sites have provided definitive evidence for burials with artifacts.[283] At Skhul, a contracted burial is associated with a pig's mandible. Qafzeh cave has *Homo sapiens* graves with ocher-stained stone tools and 70 pieces of red ochre nearby. In a "veritable cemetery," Mellars details a "clearly ceremonial burial," in which red ochre was used to stain bones, and a large pair of deer antlers were deposited on the hands of an adolescent.[284] A ritual "syntax" is displayed by this pairing of human and animal bones, the placement of tools, and the use of a coloring agent. Red ochre stained burials, as already intimated, are laden with possible symbolic meaning, such as revitalizing the body after death. Intentional ritual activity suggests a strong symbolizing capacity, and one consistent with the emergence of a complex syntax in language.

Evidence for an intensifying social life is consistent with a full language . Dating 110 - 100 Kya, from the Klasies River region of South Africa, cultural continuity is evidenced by 20 meters of deposits, mostly hearths and middens of shells and bones. By 100 Kya, African *H. sapiens* constructed postholes and left behind a semi-circular spread of thick waste, bone, and ash.[285] More definitive identification and dating of human adornment comes from perforated shell beads at 100 Kya.[286]Last, from Qafzeh cave, broken Levallois cores are incised with parallel and overlapping lines.[287] Experts have not missed the significance of these finds, stating they mark the generation of a "symbolic referential framework" and the birth of a "symbolic culture.[288]

Let us now review the evidence in chart form:

*1/3 Explicit node (97 Kya) Qafzeh Cave/patterned markings 100-90

*1/4 Explicit node (103 Kya) Skhul/Qafzeh/Intentional burial 101- 95 Beads 100

$2^{16.9}$ ❂ Major node (122 Kya) Home range expands 130 Dart symbol c 120

First step of Globular Brain Change (Group II) c. 115-120 Kya

10% 131

*1/4 Precursor node (145 Kya) 25% 145 Perforated beads Bizmoune Cave, Morroco 142

*1/3 Precursor node (154 Kya) Hereto skull with markings 161-150

+ Mid-node (172 Kya) Molecular estimates for *Homo sapiens* 185-149 Omo-Kibish 190-100

In summary, the 7th Human Node of Human Evolution, which is the first node of another set of four, marks the appearance of *Homo sapiens* with a globular brain and significant cultural innovation. They include personal adornment, two-dimensional iconography, expanded trade networks, more defined living spaces, and ritual activity associated with death. These finds support the "invisible" beginnings of a full language capacity, which becomes more manifest over the next two nodes.

Chapter VIII

8th Human Node; 2nd Homo sapiens / Complex syntax 61 Kya

($2^{15.9}$) 17th Node of Life

With the 2nd node of *Homo sapiens* at 61 Kya, we look for a marked change in cognitive/behavioral complexity in terms of symbolic activity. Further, we also look for evidence of a novel social fusion.

With regard to the latter, we consider the heavily debated possibility of hybridization between African *Homo sapiens* and the European Neandertals. The Amud fossil (dated 60 - 53 Kya) has Neandertal traits in the shape and size of skull and several Anatomically Modern Human traits: small brow ridges, narrow zygomatic arches, a small mastoid, and a small chin.[289]Wolpoff and others have summoned a variety of morphological and cultural evidence indicating that Neandertals hybridized with African *Homo sapiens* and had a similar communication capability.[290] Recent research by DNA analysis has largely confirmed such hybridization, giving multiple lines of evidence for Neanderthal interbreeding. Up to 10% of the genome of early *Homo sapien* groups are Neanderthal in origin, declining to present-day levels of 2-4%.[291] While technically we can be seen as one species, others have argued for a braided concept of an evolutionary tree, which allows for distinct groups and hybridization.[292]

However these relationships sort out, there is not much doubt as to significant technological gains at this node. Mellars ascribes to the Middle Stone Age the appearance of six or seven of the most striking features of modern technology, often attributed to the "paleolithic revolution" of the following node.[293] Finds from South Africa include bifacially retouched tools ranging from narrow elliptical to lanceolate shapes, specialized end scrapers and burins, small geometric forms used as insets for multi-component hafted tools, and extensively shaped bone tools. From Zaire, harpoons, dated 75 Kya, near the 1/3 precursor node, provide evidence for river fishing and formal artifact production that required carving and grinding.[294]

The precursor period presents other strong signs for symbolic activity. One of the most prolific sites at this node is Blombos cave from South Africa, with the earliest bone tools dated 85-77 Kya, between the mid-node and 1/3 precursor node. Here, we find sophisticated use of bone tools, as well as the purposive polishing of points for no apparent functional reason.[295] Symbol making has been identified during the final archeological stage of Blombos, recently redated to 73 Kya[296] at the 1/4 precursor node. Unambiguous personal ornaments include many examples of perforated shells and an ostrich shell. On ochre slabs, abstract engravings in the shape of complex cross-hatched markings appear.[297] According to Enrico, these diamond-lattice designs are the "most ancient irrefutable evidence for symbolic behavior" of *Homo sapiens.*[298]

Rock art appears to have accentuated suggestive shapes found in natural formations. Among hills and caves sacred to the San people in Botswana, Africa, a python-shaped rock has some 300 to 400 indentations, suggestive of scales. Artifacts at the site, dated 70 Kya, near the 1/4 precursor node, include colorful flints on which red coloring was burned, indicating ritual intent.

Also, in the region of the 1/4 precursor node, human burials were paired with ornaments and/or animal bones. From Border Cave, South Africa, whose antiquity has been confirmed at 74 Kya,[299] an infant was buried with a perforated Conus shell.[300] From Uzbekistan, a

Neandertal infant burial, dated 70 Kya, was paired with a wild goat cranium. The discoverers described the tomb as circled by horns.[301] The attention placed upon the burial of children suggests strong bonds between parents and their infants.

At Regourdou, France, remarkable bear burials have been dated 70-60 Kya, between the 1/4 precursor node and the Major Node. Although doubts as to intentionality have been historically raised, recent re-evaluation of the evidence by Haydn concluded the skulls of mostly young bears were purposively placed in cists made of large slabs of pierced rock. Nearby, a Neanderthal burial placed the legs of bear bones at its feet and on top of a slab of rock on the chest, the split foreleg bone from a bear.[302] If we extrapolate from Indigenous cultures, these burials are indicative of an *animal cult.* Varied meanings can be suggested, as exampled by circumpolar cultures in which the ritual slaying of an animal helps humans to gain access to spiritual power. Rituals, as earlier noted, also make for stronger bonds within a social group, which extends to inhabitants of the spiritual world, all consistent with our expectation for an enhanced social fusion at this node.

Within 10% before the Major Node, and before modern humans arrived in Europe, we have evidence of artistic activity, likely from the Neanderthal people. Accurately dated by Uranium-Thorium methods, these include a hand stencil at 67 Kya red, rectilinear cave painting at 65 Kya, and in a cave, a limestone fin painted red, also dated to 65 Kya. Thus, symbolic activity of Neanderthal people extended to expressions in design.[303]

Further evidence for symbolic activity can be found in Neanderthal burials. At the time of the Major Node, circa 60 Kya, a juvenile child was covered with a limestone slab whose underside was marked by 18 cupules, of which 16 were paired.[304] Also at 60 Kya, a Neandertal hunter in Shanidar cave was laid to rest on a bed of pine boughs with

28 different types of flowers identified. Although this burial has been questioned, a careful reevaluation which accounts for diversity, cave location, depth, and the aged condition of the flower remains led Leroi-Gourhan to conclude that "funerary practices is indisputable here."[305] Recent skeletal finds have also been confirming of intentional human burial.[306]

Strong cultural continuity is evidenced at this node. In a number of sites, the repeated use of hearths in the same area of caves documents generational continuity. Kebara cave from Israel, for example, has continuous deposits spanning 60 Kya to 12 Kya. Shanidar Cave in Iraq, dated circa 60 Kya, and La Ferrassie rock shelter in France were used for multiple burials over long periods of time, indicating mortuary centers and generational continuity.[307]

The 1/4 and 1/3 explicit nodes at 51.4 and 48.5 Kya presents us with rich symbolic finds. From Western Cape Province in South Africa, 200 incised ostrich fragments, dated 65-55 Kya, bear a variety of deliberate designs.[308] From the Golan Heights in the Middle East, a flat cortex flint plate, dated 54 Kya, bears a complex tent-like design, characterized by Marshack at the time as the earliest abstract art.[309] This artifact has been described as depicting a gestalt or composition (a preconceived plan of a design) and a clear intention to cover the entire available surface.

The earliest evidence for narrative cave art has recently surfaced from the Indonesia island of Suawesi. With parts dated to 51.2 Kya, 48 Kya and 45.5 Kya, a scene is depicted with the viewer can infer interactions between humans and animals. Remarkably, the oldest scene depicts humans, handprints, and wild pigs in a compositional arrangement.[310]

Two more examples at 1/3 log distance will close out the explicit period. From a deep cave in Bruniquel France, a ring of broken stalagmites, with a minimum age of 46.7 Kya but generally dated 50

Kya, appear to demarcate a ritual space.[311] From the Olay region of South Australia, an oval motif and curvilinear motif have been dated to 42 Kya and 43 Kya, respectively.[312] In sum, the explicit period provides confirming finds for complex symbolic expression: abstract designs, narrative cave art, elaborate burials, and ritual spaces in deep caves.

Let us now review the evidence in chart form:

+ Mid-node (43 Kya) Oval, curvilinear designs 43-42 Kya

*1/3 Explicit node (48.5 Kya) Stalagmite ring 47.6 Projectiles 50-40

*1/4 Explicit node (51.4 Kya) Tent-like design 54 Flower burial 50
Narrative cave art/ Indonesia/ Sulawesi cave 51.2 K

*10% 57

$2^{15.9}$ ❂ Major Node (61 Kya) Bear burials 70-60 Continuous hearths 60-12

10% 65.5 Red rectilinear cave art 65 Hand Stencil 67 / likely Neanderthal

*1/4 Precursor node (73 Kya) Python marks 70 Cross-hatch design 73

*1/3 Precursor node (77 Kya) Infant burial with shell 76 Harpoons 75

+ Mid-node (86 Kya) Fine bone tools 90-75

In sum, a variety of finds support a logarithmic leap in symbolic capacity, with the presence of a complex syntax. These include several examples of abstract designs, and pairing burials with personal objects, animal bones, and flowers. If this node provides increasingly clear evidence for symbolic activity, we expect a yet more full elaboration by the next node.

Chapter IX

9th Human Node; *3rd Homo sapiens* Complex syntax/ Art & Notation 30.6 Kya

$(2^{14.9})$ 18th Node of Life

At the 3rd node for *Homo sapiens,* we come to the Upper Paleolithic and a flowering within the symbolic-cultural sphere. We will also look for a heightening of social fusion and resilience demonstrated by technological advances and/or geographic spread. With regard to the latter, by 40 Kya modern humans had spread into Europe and the mid-latitudes of Asia, after which they blended with various archaic populations.[313]By 30 Kya, humans were making the ultimate land conquest, penetrating deep into the arctic regions of Europe and Siberia.[314]

Let us consider the archeological finds. Near the region of 1/3 precursor node (39 Kya), the Aurignacian culture is traditionally dated 40 Kya, coinciding with the advent of modern *Homo sapiens* in Europe. In recently recalibrated carbon years, the dispersal of Aurignacian type tools across central and western Europe may have occurred even earlier, closer to the mid-node, between 46 and 41 Kya.[315]An advanced Neandertal tool complex known as Chatelperronian, equivalent in many ways to the Aurignacian, goes back to at least 38 Kya, and possibly to 42 Kya.[316]

The Aurignacian period gained its name from a tool technology in Europe which was worked from prepared cores and then made into thin, long *blades*, at least twice as long as wide and whose retouching was very fine. The beauty of the blade technology is that, from a core, a dozen standardized pieces could be produced, which could then be made into various specialized pieces ranging from engravers to spear barbs.

The area of the 1/3 and 1/4 precursor nodes presents rich artistic finds. In what Conrad calls a "cultural fluorescence," the first naturalistic 3-D depictions of animal carvings appeared at 40 Kya in the Danube region.[317] Astonishingly evidence for musical instruments appears, "reed pipes" dated 37.8 - 29.6 Kya. The Geisswenklosterle flute, at 36.8 Kya, was carved from the radius of a swan bone and possessed elaborate scales and notches which gave precise musical effects. As the flute is a complex instrument, we assume an even earlier date for the first musical devices, which were likely to be percussive.[318]

The earliest paleolithic figurine, dated to at least 35,000 years ago, near the 1/4 precursor node, has been found in the Swabian Jura region, from Hohle Fels Cave in Germany. Carved from mammoth ivory, this headless female possessed oversized breasts and exaggerated genitals. Striations mark the figurine, in particular the arms and abdomen. This find has forced a reevaluation of the history of sculpture, as it is found 5,000 years earlier than the well-known Venus figurines of the Gravettian era.[319]

Recent finds in Italy and Germany indicate that cave art was more widespread and much earlier than previously thought. Between the 1/4 and 1/10 precursor nodes (34 - 32 Kya), incised stone slabs have been found in a cave near the Alps. Depictions include a horned anthropomorph holding a ritual object and an indefinite quadruped.

While symbolic expressions on stone dominate our attention, other expressive mediums were also present. Dating back to 34 Kya, woven fibers have been found in a Georgia cave at the foothills of the Caucasus. The wild flax fibers were spun, twisted, and dyed. Multiple uses were likely, such as the making of cords for hafting tools, cloth for garments, and strands for baskets.[320]

Dated 33 Kya, within 10% of the Major Node, we return to Hohls Fels Cave in Germany, where three remarkable miniatures figurines have been found: a horse, a diving water-bird, and a half lion/half human figurine.[321] The hybrid motif of animal-human may be related to spiritual activity, involving ritual identification with an animal. In this regard, it could be associated with adolescent passage, with hybrid animality signaling the entrance into another phase or plane of being.

We now consider in more detail this early highlight of the Node, with finds from three major sites in Swabian Jura. Of the array of objects found and which are most commonly made of ivory, about 50% bear symbolic markings, of which figurines are the most heavily marked. The various patterns are composed mostly of dots, lines and notches, and present a challenge to decipher. The beautiful and largely natural looking figures include a human with upraised arms, the rotund feminine figurine previously mentioned, a mammal, a feline, often overlaid with complex abstract markings, which at times appears to coat the figures with symbolism. A clear preference exists for marking the left side of the figure; women most often have line patterns, while therianthropes (hybrid human and animal forms) mostly have notches and dot patterns. Thus, a culturally encoded symbolism is present.[322] The portrayal of beauty with the symbolic world should also not go unnoted. When humans created symbolic objects, an aesthetic was present, which served to focus attention, create wonder, and become a tool for exploration.

The dating of the finds of three sites, which hold the symbolic finds, generally ranges from the 1/3 precursor node to 10% past the major node. The mean of a table of dates, however, is 35 Kya just past the 10% precursor node, with the most common dates found (32,33,34,35 Kya) revolving right around the 10% precursor node. The impression gained from this beautiful array of finds is that a symbolic world and understanding is permeating human culture at this time.

More generally at this time, we can document the routine appearance of tally-like marks between 35 and 30 Kya (from the one-fourth precursor node to the Major Node) in sites such as Abri Blanchard, France. Engraved rows of marks on bone and ivory have been carefully analyzed microscopically and been interpreted by Alexander Marshack as a lunar calendar for ritual events.[323] Although such interpretations are disputed, a notation system of some kind appears to be present, making for a type of "artificial memory." [324]

Paintings on a large scale are dated to this general time, with the surprising early find of a fully decorated, deep cave, Chauvet-Pon d'Arc in France.[325] Recent dating of most of the charcoal drawings date from 32 to 30 Kya, right in the region of the Major Node.[326] The cave images present fully modeled figures drawn with expertise, with a planned and decided upon theme of impressive non-hunted animals. The hunt by predatory animals, such as the large panel of lions, has been proposed as the "central preoccupation" of the artists. To this we can add an explicit feminine dimension. In the deep region of the cave, a hybrid figure—part bovine and part human—is integrated with a vulva representation. The organization of figures in "complex compositions, conceived as narrative elements" again points to spatially coded relationships or "symbolic complementarities."[327]

Last and not of less importance in Chauvet, a bear cult is manifest, as dozens of bear skulls have been found in a restricted area, of which one was conspicuously placed on a rock slab. In a description by the French archaeologists,

> The Skull Room presents the astonishing singularity of containing more than fifty bear skulls, within a perimeter of about ten meters at the foot of the large rock block, on which also rests a bear skull. [328]

Thus, it would appear that Chauvet cave provides evidence for a symbolic system which incorporates passage within an underground setting.

At this node, as we sense from the previous example, elaborate and heavily ritualized burials become more manifest. At 30.4 Kya, right at the Major Node, pit structures in Siberia with stone walls and floors within a larger circular structure have been identified. Within the pits, a wolf's skull and a complete set of horse bones were interred. This find has been interpreted as a ritual burial of a predator's head and a sacrificial offering of a horse.[329] Here, we can sense the presence of an archaic "temple" with a center and periphery.

A little later at 29-28 Kya and approaching the 1/10 explicit node, the so-called Red Lady of Paviland comprises a fairly complete skeleton, which has now been determined to be a male. His bones are dyed red and accompanied by jewelry made of ivory and sea shells and a mammoth skull.[330]

From Sunghir Russia, a complex ritual burial has been unearthed and dated to 28 Kya. A boy and girl child were buried head to head, layered with red ocher, and accompanied by rich grave goods. The latter includes thousands of ivory beds, long spears of straightened mammoth tusks, and perforated arctic fox canines. This excess implies a strong sense of community and the social importance of children.

The explicit period presents us with a plethora of decorated caves. In France two caves filled with designs have recently been discovered and dated to 27 Kya, within 10% of the Major Node. Cosquer, the cave beneath the sea, possesses an array of animal depictions and a variety of hand imprints. Both positive and negative hand impresses, which often have missing fingers, have been variously interpreted: youths at an initiation, ritual mutilation, gestural symbols related to language, and even an incipient mathematics.[331] In the second cave from the Vilhonneur grotto, the cave walls and artistic lines are integrated to make for the oldest known portrait. The presence of human hand imprints and images of a human face imply a strong sense of personal identity.

The 1/4 and 1/3 explicit nodes, 25.7 / 24.3 Kya, generally frames the Gravettian period and presents many of the most famous painted caves. Although rock art had become a worldwide phenomenon by 25 Kya, we focus on Europe as they have been well documented and dense. Cougnac is dated 25.1 - 23.6 Kya, and Perch-Merle, with its famous Chapel of Mammoths, is dated 24.6 Kya. Imagery from the European Gravettian presents several "striking new elements," including large numbers of human hand prints and hundreds of representations of humans, which are almost always pregnant women.[332]

One of the most famous female figurines, the Venus of Willendorf, was created during this general time frame, from 26 to 24 Kya. Female figurines, often rotund, have been generally interpreted as being associated with fertility, birthing, maternal care, and plenty. A passage experience into adulthood or shamanic initiation in which the female serves as a spiritual midwife can also be considered. Finally, geometric patterns, most often parallel and crosscut designs, are found on such figurines, and are possibly related to personal and social identity.[333] Whatever the specific meanings, we find rich, symbolic associations paired with the feminine.

From 27 to 23 Kya and during the explicit period, burials incorporate the heavy use of red ochre, and the bones of large herbivores such as mammoth and bison. Most human skeletons have pathological features, indicating these individuals experienced considerable pain for much of their life. Although this may simply document that hardship was common at this time, handicapped persons may have been viewed as special with shamanic powers.[334]

The explicit period provides a repeated association between burial and art. In the Grotto de Cussac, a number of burials, dated 25 Kya, lie at the base of an engraved wooly mammoth. A little later at 24.5 Kya, the famous "hybrid" child (with modern and Neandertal traits) was buried with pierced shells and red ocher.[335] From Bruno II of the Czech Republic, a male human burial, dated 23.7 Kya, was accompanied by a multi-component ivory sculpture. Burial of the dead, with aesthetic works of symbolic import, marks the start of a long human tradition of expressing love to the deceased.

All these art objects and burials, we have thus far focused upon, are part of a larger way of life. In central Europe the pattern of human settlements amazes us in terms of complexity and continuity. The open-air settlement of Dolni Vestonice in the Czech republic has several cultural layers, dating between 29 to 22.2 Kya. Some 100-120 people lived in oval bone and timber winter huts up to 15 meters long and 9 meters wide, with hide roofs. This site has an abundance of the first fired ceramics, as well as woven fabrics which utilized 13 types of knots, and probably a loom. Small fired ceramics of animals (mostly carnivores) and humans (mostly women) were exploded in kilns, perhaps ritually.[336] To this we add the oldest known baked clay figure at 26 Kya, a female with exaggerated breasts and hips. Last, we highlight a triple burial of a woman and two men who were possibly related. The surrounding area was sprinkled with red ochre. Beads and pierced animal teeth were placed near the male skulls, possibly indicating elaborate headdresses. There is also evidence for the partial burning of wood over the grave.[337] Again, we sense that an elaborate ritual, paired with a symbolic system, had arisen by this node.

Situated on a promontory overlooking the Don River in the Ukraine, the Kostenki site has a long history dating back to the precursor period, at least to 38 Kya and possibly to 40-45 Kya.[338]We focus our attention, however, at the 1/3 explicit node and the time between 24 and 23 Kya, when semi-subterranean structures were gathered in an oval around a common space, along with a linear series of fireplaces. The area is riddled with female representations, often seeming to accentuate pregnancy. The jewelry and ornamentation are beautifully incised with small toothed shaped incisions, fishtail patterns, double zigzag lines, and crescent-like figures arranged in a row, which may be part of a female cult.[339] A scholar goes so far as to say that by 25 Kya the human body, in particular the female's, has become the "center of symbolic irradiation." He hypothesizes that the human body is now perceived through "procedures of transformation and transmutation."[340]

With cultural layers from a village, ceramic production, and elaborate female iconography, these central European sites lead us to wonder if the Neolithic era has arrived. Despite such strong intimations, trademarks of the Neolithic, such as pottery and domesticated plants and animals, are absent. Starting at 23 Kya and near the mid-node (21.6), however, wild grasses and other plants were utilized as staples during the so-called "broad spectrum revolution," in which novel foods were accessed to overcome shortages.[341]

We can now review the evidence in chart form.

+ Mid-node 21.6 Kya Broad spectrum revolution 23

*1/3 Explicit node (24.3 Kya) Kostenki 24-23 Dolni Vestonice 29-22

*1/4 Explicit node (25.7 Kya) Venus of Willendorf 26-24 Sewing 26

*10% Log distance (28.5) Double child burial 28

$2^{14.9}$ ❂ Major Node (30.6 Kya) Pit structures/ animal internments 30.6

Chauvet charcoal drawings 32/30

*10% Log distance (32.8 Kya) Lion-human hybrid 33 Chauvet 2nd occupation 32.4

Marked objects/figurines/Swabian Jura (~33) Wild flax fibers 34

*1/4 Precursor node (36 Kya) Notational system 35-30 Flutes 36
Early female Figurine 35
*1/3 Precursor node (39 Kya) Animal figurines 40
+ Mid-node (43 Kya) Aurignacian 46/40 Chatelperronian 42-38

In summary, the 3rd node of *Homo sapiens* presents a cultural flowering that amazes us at every turn. It includes the 2-D naturalistic images preserved in galleries of deep cave art, female and hybrid sculptures, and highly elaborate burials. If we add musical instruments and a notational system, a strong case can be made for a complex, syntactical awareness. No doubt this awareness integrated the frequently depicted animal and female forms in an archetypal way.

As we move into the explicit period, settled sites go beyond our standard sense of the paleolithic and grant us a strong intimation for the future. We are now ready for the last node in this set of four, which presents yet a more complete flowering of earlier impulses and which doubles as a new invisible beginning and quantum change.

Chapter X

10th Human Node; 1st Node of Civilization/ Early Neolithic/ 15.3 Kya

($2^{13.9}$) 19th Node of Life

With the 4th node and expectation for a quantum change, we might wonder what novel development could occur in a time typically noted for its paleolithic art. Clearly, we are out of range of significant morphological change, although from 10,000 years ago or so that the human face is shrinking some 1-2% every 1000 years, and the skull is becoming rounder.[342] We must turn to the cultural sphere where we posit the start of the Neolithic era. Specifically, we find more settled sites, cultivation of plants, elaborate iconography, and symbolic systems which led to writing. With large settled villages and proto-forms of writing, we posit the earliest stages of what is termed civilization. A sense of generational continuity and greater community bonding is evidenced by large villages which are built up into tells. Some sites are accompanied by megalithic architecture, indicating that Neolithic communities are actively structuring their own spatial-temporal world.

Let us turn to events and their placement on a logarithmic scale. In the precursor period, cave art in Europe continues with the Solutrean culture, dated 22-18 Kya. Animals thematically dominate, with the new stylistic trait of "deformation," especially horses portrayed with "heavy voluminous bodies, short legs, and oftentimes tiny heads."[343] In the general surround of the Major Node (15.3 Kya), the European paleolithic comes to full flower with the Magdalenian period, dated 18-11 Kya. Art becomes dense with an explosion of rock art and antler/ bone work comprising some 80% of all known Upper Paleolithic art.

[344]At the 1/10 precursor node, there is not much doubt as to the high level of artistic activity expressed in caves, example by Lascaux, dated 17 Kya, and Altamira at 16.4 - 14.3 Kya. In these underground sanctuaries, animal imagery dominates, with a focus on herd animals in the main galleries and dangerous animals in the cave's deeper portions. Other important, but rarer, themes appear to be related to life passages: birthing, initiation into adulthood or shamanic status, and death.[345]

Figurative cave art is complimented by abstract signs patterned in terms of their placement within the caves, pairings with animals, and a consistent frequency of appearance. Although specific interpretations are speculative, a strong case can be made for the presence of a complex symbolic system. Wildgen traces the origins of writings from a spiritualized paleolithic context before it took more standard form in an urban and economic setting.[346]

With regard to subsistence activities, the potential for a new way of life emerged after the last ice age ended between 20 and 15 Kya, when the climate warmed. In what has been termed the "epipaleolithic," from 23 to 10.2 Kya, permanent settlements arose which were sustained by foraging. There was a consistent selection of plant species, the herding of animals, as well as storage of food. Specialized tools, pestle and mortars, were used to grind ochre, acorns, and grasses.

An early example of such settlements is from Ohalo II in Israel, dated 20 Kya, near the 1/3 precursor node. This site exhibits a cluster of six huts occupied throughout the year, along with large quantities of charred plant food. Grinding stones helped prepare hard-seeded foods, such as acorns and grasses, and there was a burial among the houses. [347] The picture of a hunter-gatherer culture following animal herds on a seasonal round is giving way to a sedentary life dependent on specialized, consistent resources.

Likewise, in Central Europe, settlements were maintained even during the winter. An outstanding example is Mezhririch on the

Dnieper River, dated 18-14 Kya between the 1/4 precursor node and the Major Node. This "full-fledged settlement," with five buildings made of mammoth bone, is a "foreshadowing of sedentary societies." [348] Bone needles at the site indicate that people sewed. A wealth of cultural artifacts has been found, including complex designs on mammoth bone, jewelry of ivory, shell, and animal teeth, as well as figurines made of bone, ivory, and amber. Thus, within the paleolithic flowering, there are seeds of the Neolithic.

At the Major Node, 15.3 Kya, finds appear usually associated with the Neolithic. Although sustained by hunting, fishing, and gathering, the Jomon culture in Japan had large settlements covering tens of thousands of square meters.[349] In caves at Jomon sites, the world's first pottery, dated 15.3–13.7 Kya, was built with coils of clay, making for distinctive cord-like markings.[350] Early pottery was used for storage and transport, as well as accompanying burials and ritual ceremonies. Chinese pottery is also recorded from this general time, 15.5 - 11.6 Kya, and it's believed likely that the early pottery of Japan likely was the result of diffusion from mainland Eurasia.

In the region of the Major Node (15 Kya), dramatic environmental changes from global warming contributed to the vanishing of Magdalenian cultures. More rich, localized resources, such as coastal resources, brought a new life style termed *affluent foraging*. Signs of plant domestication appear in South Korea, with quasi-rice dated 17.3 Kya and ancient rice at 14.8 Kya.[351]The earliest rice from China dates to 12-11 Kya. The first domestication of animals, the dog, occurs at this time. The burial of a dog in Germany has been associated with a double human burial dated 14 Kya. Between 14-12 kya, canine domestication became a world-wide phenomenon.[352]

By the explicit period, we look for clear signs of the Neolithic. The earliest and best documented sites are from the Natufian culture of the Near East, dated 15-12 Kya calibrated years—between the Major

Node and the 1/4 explicit node. In this region, we can posit the rise of *sedentism*, a steady year-round occupation of a site, by 14.5 Kya. At this time, villages appear with stone foundation houses that generally were populated with about 60 people.[353] The presence of larger groups of people in a settled place for a long period, along with social ranking, is a prerequisite for civilization as we know it.

Most scholars place the threshold for the Neolithic during the Natufian at 13 Kya, near the 1/4 explicit node. At this time, we have "clear evidence for agricultural origins." Tools for *cultivation* (which precedes domestication) include: milling stones, mortars, pestles, and sickle blades which exhibit a characteristic polish from extensive harvesting.[354] Dense stands of wheat-like grasses naturally occurred, and human interaction gradually led to the selection of better types that led to agriculture. By 12 Kya, close to the 1/3 explicit node, deliberate farming based on rye, einkorn and lentils appeared at Abu Hureyra in Western Mesopotamia. With cultivation, whether of wild or recently domesticated plants, extensive collaborative effort with some form of hierarchy can be envisioned.

Agriculture has traditionally been considered the major change characterizing the Neolithic. Interestingly, agriculture in the Near East first occurred in areas where resources were rich. Perhaps a permissive environment encouraged larger groupings, which in turn led to experimentation and the domestication of plants and animals. A dramatic and stressful climatic change, however, occurred shortly before domestication and may have provided an impetus for change. [355]

Burials became more structured, and the practice of creating elaborate grave goods intensified. For the first time, "hunter-gatherers" buried their dead in organized cemeteries around living sites. Burials commonly included personal adornment, such as capes, bracelets, hair ornaments, and garters embroidered with dentalium shells. In Natufian cemeteries, dentalium seashells were confined to a few burials, indicating a social hierarchy. In the late Natufian, a preference for using green colored stones with personal ornaments emerged, which may have been a fertility charm.[356]

During this time, an "outburst of symbolic expression" employed meander patterns and zoomorphic imagery in three dimensions. Long distance trade included varied materials: basalt, red ochre, and dentalium shell beads.[357] We have clearly entered a novel zone of change that goes beyond the standard view of the paleolithic.

Evidence for the Neolithic strengthens as we progress into the explicit period. With recently recalibrated dates of 12.34 - 11.3 Kya, between the 1/4 explicit node and the mid-node, Zawi Chemi, the site of Shanidar, gives evidence for the selective slaughtering of immature sheep and goats, indicating domestication. About 11.6 Kya, village life and large populations emerged abruptly in SW Asia.[358] Hallan Cemi, Turkey (12-10.5 Kya) lasted centuries and had a larger, more heavily constructed building which has traits of a shrine. The skull and horns of a wild bull hanging on its walls prefigured a cattle cult which may have involved sacrifice.

The Pre-Pottery Neolithic period (PPNA) lasted from 11. 7 to 10.5 Kya, calibrated, which is between the 1/3 explicit node and the Mid-node. More widespread than originally thought and with over a dozen sites, the PPNA in the Levant marks the formal transition to the Neolithic before the advent of pottery. At this time, we can speak of an agricultural economy with settled farming life based on cultivating cereals and pulses along the Levantine corridor.[359] The morphological evidence for domestication of plants and animals, however, remains weak at this time. This is to be expected, as morphological change is the end result of a long process in which species are used in the domestic sphere.[360]

About 11.8 to 10 Kya, also between the 1/3 explicit node and the Mid-node, we can place the remarkable site of Gobekli Tepe, whose name means "hill of the navel." On the eastern part of the Fertile Crescent, the world's first monumental architecture can be interpreted as temples of a regional center. In six separate structures, oval enclosures are formed by large T-shaped pillars weighing up to ten tons and decorated with a variety of animal reliefs. Two twin monoliths in the centers of each structure are especially large and well designed.[361] As some stones are clearly anthropomorphic with etched hands and arms, plausible interpretations include deity or ancestral representations. One interesting figure incised into the floor of a temple appears to be a female birthing twins. Although there are no signs of plant or animal domestication, there was ritual feasting. In sum, the site of Gobekli Tepe can be said to represent an active "domestication of the landscape," which required a level of social organizing more typically found in city-states.

Between 12 and 11. 5 Kya, a general "revolution in symbols" has been posited which focuses on the bull and the female.[362] Supporting this is the placement of bull crania, where horns are buried inside houses and included as features in many Neolithic temples. Second, numerous female figurines are often portrayed in seated positions with their buttocks emphasized.[363] For Cauvin, the "goddess" and the bull emerged as a type of primeval moiety of Neolithic society.

Last, near the mid-node at 11 Kya, the site of Abu Hureyra obtained a population that reached into the thousands and was occupied for thousands of years. This site details the transition to full-fledged agriculture with the domestication of goats and sheep.[364] It is likely that domestication of animals was integrated into the community's symbolic system. The raising and fattening of wild cattle had both a ritual and sociological purpose: feasting from which chiefs gained status and power. Feasting may also have served to build alliances between social groups, gods, and ancestors, as well as to mobilize labor.[365]

The Neolithic is often viewed as a complexification and "advance" in civilization. But a sedentary lifestyle also represents a type of simplification, as groups had to deal with only one ecosystem and more regular surroundings. The Neolithic revolution brought costs as well. Several paleopathological studies document an initial, general decline in the quality and length of human life.[366] Although overall consistency of food supply improved, agriculture brought periods of famine, as groups relied on more singular sources of food. Last, when humans upturned the soil and lived closer to animals, they became exposed to novel diseases. Such observations are consistent with our recognition that vulnerability often accompanies transition periods.

We now review the evidence in chart form.

+ Mid-node (10.8 Kya) PPNA 11.7-10.5 Settled farming life Abu Hureyra 11 Gobelki Tepe/ megaliths11.8-10

* 1/3 Explicit node (12.1 Kya) Shanidar 12.34-11.3 Domestication 12

* 1/4 Explicit node (12.85 Kya) Dog burials 14-12 Cultivation 13 10% 14.3

$2^{13.9}$ ❂ Major Node (15.3 Kya) Natufian 15-12 Rice 15 Jomon Pottery 15.3- 13.7

*10% Log distance(16.4) Lascaux 17 Altamira 16.4 Pottery 16.5 -14.9

*1/4 Precursor node (18.2 Kya) Magdalenian era 18-11

*1/3 Precursor node (19.3 Kya) Ohalo II 20

+ Mid-node 21.6 Kya Solutrean culture 22-18

In summary, this node presents the final flourish of the European paleolithic, a strong Neolithic current in Southwest Asia, and tell-tell signs of the Neolithic in the Near East. Indicators for the Neolithic include settled villages, cultivation of plants, large set-aside cemeteries, and elaborate symbolism. In the European paleolithic, a symbolic notation arose, an abstraction which intensified in the Neolithic and prefigured writing. Settled sites, which continued for generations and have notational systems, mark the start of a generational and an incipient sense of historical time. If we were to continue this charting to the beginnings of civilization, we would find more sure confirmation in the ensuing nodes. For now, however, we will call to an end our journey and see what we can conclude.

Conclusion

We have journeyed far from the first bipedal hominin to the advent of a Neolithic level of civilization. With 10 nodes at regular logarithmic intervals, the spectrum of human evolution went from ostensible morphological changes to more subtle, yet no less significant, changes in the cultural domain. Our ability to map events with a logarithmic scale has confirmed the general expectation that evolutionary events are accelerating as we come to the present. Just as importantly, the arrangement of events around nodes at logarithmic distances has proved to be a useful organizational tool. Additionally, we identified a sub-pattern in which events can be viewed as unfolding in sets of fours, in which the fourth node features a quantum change. Last, the evolutionary future often arose as a subtle sub-current amid more dominant expressions. In sum, a logarithmic scale provided us with a novel lens to view evolutionary events, one that revealed new pattern and detail in the temporal domain.

We might ask: what benefit is gained from our study? In learning about our evolutionary past, we maintain we are better able to preserve our evolutionary future. If the past serves as any witness, the future is often present as a hidden undercurrent, a fragile beginning, which needs protection and cultivation. We are invited to consider that minority trends, which are not yet part of the cultural mainstream, and in all likelihood are met with resistance and rejection, may be vital for our survival. If cultural evolution is still ongoing in humanity—and we have no reason to think that it has stopped—some individuals and/or groups exist are already undergoing changes which will be critical for humanity's future.

We can attempt to apply this to persons in our culture, who are in crisis and dealing with an excess of symbolic reality. Persons undergoing a break from normal reality often experience an archetypal depth, which involves death/rebirth imagery and a titanic struggle between good and evil. Such experiences are often overwhelming for people undergoing them, and others usually find it difficult to understand and handle them. Yet, such experiences may be the beginnings of an inner transformation and an emerging consciousness which may be helpful to the world. The stories of many patient advocates in the field, as well as some psychiatrists such as John Weir Perry, Loren Mosher, and Curtis Adams have advanced such theories.[367] Another illustration is provided by Williams and Taylor, who considered the evolutionary value of persons diagnosed with Attention Deficit Disorder. As one of the key features of ADD is diversity, they hypothesize this trait is beneficial to the group as a whole. In their view, unpredictability optimizes results for the group.[368] Our society, focused on the short-term attempts to return persons to "fitness" quickly, will short-circuit natural healing processes. In our lack of understanding of such experiences and their potential evolutionary role, we risk short circuiting our own future.

Finally, we ask what does this logarithmic study portend for the *future* of humanity? As we've seen, the evolutionary flow of time leads us to *history,* which demands its own temporal accounting. Ortega y Gasset suggested we use *indications,* 15 year increments to map the flow of generations.[369] The advent of generational historical time follows its own set of rules, in which history unfolds with regular phasing.[370]

In recent years, history has accelerated so much there may be yet a third type of time. In *global time,* different layers of time and the full depth of history are becoming conscious to humanity. The internet is an outstanding example of how information is both instantly drawn upon and spread across the globe. Thus, historical time appears to be entering a new temporal sense, which would require its own exposition. Suffice it to say that this new emergent consciousness involves a growing awareness of our own evolutionary depth. That is, to become fully human in this new age, we need to become conscious of our early prehistoric cultural roots, as well as the variety of cultures which exist across the globe. Put another way, in order to understand our own humanity, we need a sense of our deep origins, of the context that birthed us.[371]

Ultimately we are called to be *symbolic beings*, who are conscious of our deep past and open to the novelty of our evolutionary future. One of the leading American philosophers, Charles Pierce, held that *love* guided this evolutionary process. Consistent with this supportive foundation, this work calls us to protect fragile, emergent humanity and support the fragile change being continually birthed in humanity. It has traditionally been the place of religions to help guide such change processes, but an array of therapeutic modes can also be supportive. Increasingly, People are turning to alternative and Indigenous spiritualities, reviving traditions with deep roots in our past.

In sum, our humanity calls us to become conscious of our capacity to evolve and to be open to the evolutionary future, for this has been a consistent trait in our biological/cultural being. When enough of humanity has been brought into a new state of awareness and enjoyed a period of stability, then others will be called to endure an instability that calls for yet more growth. Our challenge is to grow along with evolving consciousness of humanity and to let each generation continue this process in which we ourselves are surpassed.

Summary Temporal Chart
Ten Nodes of Symbolizing Humanity

10.5. • 10.8 ($2^{13.4}$/18.5 log) Neolithic Early Aceramic/ Intensive Farming

Göbekli Tepe/Megaliths 11.5-10 Abu Hureyra 11

33% 12.1 Shanidar 12.34-11.3 Domestication 12

25% 12.85 Cultivation 13 Dog burials 14-12

10% 14.3

10. ❂ <u>15.2Kya</u> ($2^{13.9}$/19 log) Jomon Pottery

10% Log distance(16.4) Lascaux 17 Altamira 16.4 Pottery 16.5 -14.9

25% (18.2 Kya) Magdalenian era 18-11

33% (19.3 Kya) Ohalo II 20

+9.5 • 21.6 ($2^{14.4}$) Solutrean culture 22-18

33% 24.3

25% 25.7 Perche Mal 25

15% 27.5 Cosquer 27

Seven painted stone plaques /Earliest African Apollo 1 cave Nambia 27

Red ochre pigments used

10% 28.5 Gravettian 28

9. ❂ 30.6 Kya ($2^{14.9}$/18th log) Complex syntax/ Art & Notation

Venus of Willendorf 32

10% 33.8 Chauvet 32.4 German objects/figurines/Swabian Jura (~33)

25% 36.4 Music 36-20 Opercula Pore-epic cave, Ethiopia 37 k

Chauvet 36/calibrated Grotto des Gorges carving 36.2

33% 38.5 Vogelherd 40.5-31

+8.5 • 43Kya ($2^{15.4}$/17.5log) Ochre processing tools 45-40

Digging sticks with perforated stones 44K Ochre mining Lion Cavern Swazlland 43 K

33% 48.5 Kya Upper Paleo blade production replaces Levallois 45-47Kya

25% 51.4 K Narrative cave art/ Indonesia/ Sulawesi cave 51.2 K

Neander Cupule Burial 50K

Definite Meander 54 Bear Cult 50K Blade Tools revolution 50 K

Complex I Goda Buticha Ethiopia 55-47

10% 57 Kya Earliest art Europe/ Neanderthal 57 plus/minus 3

8. ❂ 61 Kya ($2^{15.9}$/17th log) *2nd Node Homo sapiens* / Complex syntax

Origins of art in Africa 80-60k Standardize at Dipekoff rock shelter 60 Kya

Standardized markings on ostrich eggshell Klipdrift Shelter 65 K

10% 65.5 Apollo 11 Cave **Therianthrope painting on slab (62-59)**

Lithics miniaturize/small grain rocks/Kenay Panga ya Saidi 67 Kya

25% 72.7 Kya Skeleton with shallow area/ Border Cave 74K

Ochre on perforated shells. Blombos Cave 75 Kya

33% 77 Blomos Cave abstract design /standardized beads 77

Infant burial with shell 76kya Complex II Goda Buticha Ethiopia 81-55

16 fragments from 100 to 70K; Oldisleben Scapula incisions 80K /recurrent events

40% 86.5

+7.5. •93 Kya ($2^{16.4}$/16.5 log)

BURIAL South African Incisions 100-80K

Flake/red ochre/incisions burial at Qafzeh 100-90

99 Engraved cortex 90-100K

Definite burial *Neanderthal* 90K Ochre Burial Qafzeh 92K

25% 103 Kya Ochre clearly used many sites 110
Qafzeh Cave
Ochre processing toolkit Blomos Cave 100 K

7. ❂ 122 Kya ($2^{16.9}$/16th log) *1st Node of Homo sapiens* / Complex syntax

Multiple groups join (135-120)
10% 130
25% 145 Earliest Beads ornamental Bizmoune Cave, Morroco 142

+6.5 •161Kya ($2^{17.4}$/15.5 log) **6th Archaic** Clear Use of Ochre 162

Habitual phase of ochre use 160 to 40
20% 213 Mastery of fire/ Border Cave 200K

6. ❂ 245 Kya ($2^{17.9}$/15th log) *3rd Archaic sapiens* / Symbolizing Humanity

HABITUAL OCHRE USING /Ritual HOMININS
10% 262 Berekhat Ram Figurine c 260K
25% 291
323 Emergent phase of ochre 330-160

+5.5. •346 Kya ($2^{18.4}$/14.5 log

33% 388 Tan Tan Figurine 500-300K
25% 411 Wonderwerk art materials 500-300K
Symbolic markings Bilzingsleben, 420-350
Sima de los Heusos, Atapuerca Shaft burials 427Kya
10% 456 Blade Lithics Kapthurin Formation 465-395

5% 472 Earliest use of structural wood/logs Zambia 476 plus/minus 23

5. ❂ 489 Kya ($2^{18.9}$/14th log) *2nd Archaic sapiens*
SYMBOLIC MARKING HOMININ
/ BLADE LITHICS/RED OCHRE /MANUPORTS

Fauresmith industry ~500 Kathu Pau (500-30) Excavation 6 Wonderwerk cave

Initial phase of ochre use 500-330

Earliest evidence of ochre use in Kathua Pau South Africa 500 K

Marked Bovine tibia Kozarnika Cave, Bulgaria (Layer 12) >500 (<600)

10% 524 Marked Aquatic Shell 500-400K Bodo skull with ritual marks 523±4

25% 582

30% 602 Rapid increase in tool complexity 600K Bodo Cranium 600

H. bodeoenis (mix of *erectus* & *Homo sapiens*) c 600kya

Intensification at Wonderwerk Cave 620

+4.5 •645 Kya ($2^{19.4}$/13.5 log)

Social Learning kicks in 600K

768 Polished plank of wood Gesher Bernot Ya'aqov 780 kay

33% 777 Gesher /Levallois tools 780

Use of Fireto detoxify plants/ Gesher Benot-Yaaqov 780 k

Gesher /Levallois tools 780

25% 823 3rd Period of Tool Complexity may start as early as 800 Kya or 600K

15% 882

10% 913 Earliest *antecessor* TD6 950-900

Butchery Marks on H. antecessor 949-772

4. ❂ 978 Kya ($2^{19.9}$/13th log) *1st Archaic Homo sapiens*/ Symbolizing Humanity /

FIRE & PREPARED CORE MAKING HOMININ/ LEVALLOIS TOOLS

Stage III Rapid technological accumulation > 1 Mya

"Buia man" Statriography. 992 Paleomagnetic >1.072
Transporting raw materials for Auecheulian long distances ~ 1 Mya
10% 1.05 *Homo antecessor* 990-790
15% 1.09 Early Prepared Cores 1.1
25% 1.16 Stratum 10/Acheulian younger than 1.19

+3.5 •1.38 Mya ($2^{20.4}$ /12,5 log) Talley marks 1.4-1.3 Spheroids Ubeidiya 1.4

1.4 Mya Impose shape! 1.48 Cut Marked Hominin fossil/ Koobi Fora 1.45 Mya

35% 1.54 KNM-Wt 1500 Rib cage of H. erectus not slender
Controlled Fire at Koobi *For A. kenya* 1.5
25% 1.65 North Africa Acheulean emerges 1.67
Derived frontal lobe/brain size increase with shape 1.7 to 1.5
15% 1.76 Early Acheulean West Turkana, Kenya 1.76

1st Acheulean tools/hand axes 1.76 Kokiselei Turkana/Konso, Ethiopia
Semi-circular basalt structure 1.75

1.8 to .6 Mya 2nd Period of complexity in Tools/Acheulean 4-7 units
10% 1.83 DS (David's Site, 1.84 Ma) in Olduvai Gorge Central provisioning places
Grooved "pendant" possible palaeoart 1.87

3. ❂ 1.96 Mya ($2^{20.9}$/12th log) **ACHEULEAN/ SYMETRICAL TOOL-MAKING** *3rd Hominin Homo* family H. erectus /Ethiopia 1.95

Earliest Olduvai hominins exploit unstable environments 2.0

10% 2.1 Core with centripetal flaking/retouching 2.2

15% 2.17

25% 2.33 More advanced Oldowan tools. 2.34 *Homo habilis* 2.3 Sure dating *Homo* 2.3

33% 2.47 Early Homo 2.5-2.4

40% 2.58 Oldowan Pebble Cores 2.6 Simple Flakes Boloi Dra <Ethiopia ~2.6 Mya Substantial Brain Increase Pleistocene starts 2.6

+2.5 • 2.77 Mya ($2^{21.4}$/11.5 log) No increase in carnivory 2.8 to 1.2 Mya

35% 2.97 Oldest Oldtown tools c 3 -2.58 *Paranthropus* 2.9 -1.2

Diverse foods/mega-processing

First period 3.3-1.8 1 to 4 units of complexity

33% Explicit node (3.1 Mya) Lucy skeleton 3.1 "Face" cobble 3

25% Explicit node (3.3 Mya) *A. africanus* 3.3

Simplest Tools Bipolar percussion sequences W. Turkana Kenya ~3.3

Tool assisted behaviors 3.4 Mya

10% Log distance (3.65 Mya) *Kenyanthropus. platyops* 3.4-3.57 Footprints 3.59

2. ❂ 3.9 MYA ($2^{21.9}$/11^{th}log) **TOOL-MAKING HOMININ**

2nd Hominin Node & arboreal climbing/suspension

Australopithecus afarensis 3.8-2.9

10% Log distance (4.2 Mya) *Anamensis* 4.2 - 3.9

1/4 Precursor node (4.65 Mya) *A. Ramidus* 4.5 - 4.3

1/3 Precursor node (4.9 Mya)

+ Mid-node (5.53 Mya) *A. kadabba* 5.8 - 5.2 /Loss of canine honing

+1.5 5.53 Mya ($2^{22.4}$/10.5log) *Ardipithecus kadabba* Most samples 5.8-5.4
Other *Orrorin* specimens. 5.8

10% 5.93
15% 6.14 Single tooth? 6.3 *Orrorin* 6.1-5.7 Earliest specimen 6.1
1/3 Explicit node (6.2 Mya) Bipedal *Orrorin* 6.14 Rifting in Plate Tectonics c 6
Molar tentatively assigned to *A. kadabba* 6.3

1/4 Explicit node (6.6 Mya) *Sahelanthropus/bipedal* 1st hominin ~ 7 (6.8–7.2)
10% 7,3 Hominins in East Mediterranean 9.6-7.2

1. ❂ 7.8 Mya ($2^{22.9}$/11th log) **BIPEDAL -GESTURING HOMININ** 1st Hominin Node
10% log distance (8.4 Mya) *Oreopithecus/*bipedal shuffling 8.2
*Anadololuvius/*Hominine from Turkey 8.7
1/4 Precursor node (9.3 Mya) *Ouranopithecus* 9.5–9.3 *Dryopithecus* 9.5
Hominins in East Mediterranean 9.6-7. *S. Kiptalami* 9.6
1/3 Precursor node (9.86 Mya) *Nakalipithecus nakayamai 9.9–9.8*
+midnode: 11.06 Late Miocene (11.65- 5.33)
Danuvius guggenmosi /11.62/ *Model for common ancestor of Great Apes & Humans*

Endnotes

[1] Craig, William 2021. When Did We Become Human? Excerpt from *Quest of the Historical Adam*. Quoted in *Peaceful Science*.

[2] Watts, I. Chazan, M. & Wilkins, Jayne 2016. Early evidence for Brilliant Ritualized Display: Specularite Use in the Northern Cape (South Africa) between ~500 and ~300 ka. Current Anthropology 57 (1).

[3] Susko, Michael 2020. *Ten Pulses of Evolution and the Surprising Nature of Evolutionary Time*. AllrOneofUs Publishing.

[4] Bobe, R. & Wood, B. 2021. Estimating origination times from the early hominin record. *Evolutionary Anthropology*, DOI. 10.1002.

[5] Jacques Prieur, et al. 2020. The origins of gestures and language: history, current advances and proposed theories. *Biological Reviews* 95, 531-554.

[6]Dingyu Chung 2018. “Language Neuromechanics: The Human Biological-Language Evolution, (See abstract for quote.) Journal of Behavioral and Brain Science, 8. P. 447-472.

[7] Simpson, SW et al. 2015 Late Miocene hominin teeth from the Gona paleoanthropological research project area, Afar Ethiopia. *J. Human Evolution*, 81:68-82.

[8] Haile-Selassie, Y. 2001. Late Miocene hominids from the Middle Awash, Ethiopia, *Nature* 412:178-181.

[9] Haile-Selassie, Y., Suwa, G., White, T. (2004). Late Miocene teeth from Middle Awash, Ethiopia, and early hominid dental evolution. *Science* 303: 1503-1505.

[10] Sanders, Robert 2003. New Ethiopian fossils are from 6-million-year-old hominid living just after split from chimpanzees. (Tim White quoted). UC Berkeley News.

[11] Tanner, N. & Zihlman. 1976. Women in evolution. Part I: Innovation and selection in human origins. *Signs: Journal of Women in Culture and Society*, 585-608.

[12] Lovejoy, C. 1981. The Origin of Man, *Science* 211:341- 350.

[13] Semaw1, S., Simpson, S., Quade, J., Renne, P. Butler, R., McIntosh, W. Levin, N., Dominguez-Rodrigo, M., Rogers, M.. 2005 Early Pliocene hominids from Gona, Ethiopia, *Nature* 433: 301-305.

[14] Patricia Rice & Norah Moloney 2008. *Biological Anthropology and Prehistory: Exploring Our Human Ancestory*. 2nd edition. (see section on First Potential Hominids, p. 125) Routledge.

[15] Campbell, B., Loy, J., Cruz-Uribe, K. 2005. *Humankind Emerging, 9th edition* (p. 207) Boston: Pearson, Allyn & Bacon.

[16] Begun, D. 2004. The earliest hominin – Is less more? *Science* 303: 1478-1480.

[17] Scott W. Simpson, Naomi E. Levin, et al. 2019. Ardipithecus ramidus postcrania from the Gona Project area, Afar Regional State, Ethiopia. Journal of Human Evolution 129; 1-45.

[18] Hanson, B. 2009. Light on the origin of man. *Science* 236:5949.

[19] Pickford, M. 2002. Palaeoenvironments and hominoid evolution. *Zeitschrift fur Morphologie und Anthropologie* 83(2-3): 337-348.

[20] White, T. et al. 2006. Asa Issie, Aramis and the origin of *Australopithecus*. *Nature* 440: 883-889.

[21] White, T. et al. 2006. Asa Issie, Aramis and the origin of Australopithecus. *Nature* 440: 883-889.

[22] Ward C, Leakey, M. & Walker, A. 1999. The new hominid species *Australopithecus anamensis. Evolutionary Anthropology* 7: 197-205.

[23] Plavcan, J. Ward, C. & Paulus, F. 2009. Estimating canine tooth crown height in early Australopithecus. *Journal of Human Evolution* 57: 2-10.

[24] Vaneechoutte, Mario, et al. Sept. 2023. Have We Been Barking up the Wrong Ancestral Tree? Australopithecines are Probably Not Our Ancestors. *SciePublish.*

[25] Hammond, Ashley & Carol Ward. 2013. See Summary of Chapter 23 "Australopithecus and Kenyanthropus, "in *A Companion to Paleanthropology, 1st edition.* Blackwell Publishing LTD.

[26] Robert Boyd, Joan Silk & Kevin Langergraber. 2024. (See pp. 252-253) in *How Humans Evolved 10th edition.* W.W. Norton & Company.

[27] Stamos, P. & Alemseged, Z. 2023 Hominin locomotion and evolution in the Late Miocene to Late Pliocene. Journal of Human Evolution, 178.

[28] Hunt, K. 1998. Ecological morphology of *Australopithecus afarensis.* In Strasser et al. (eds) *Primate Locomotion.* 397-418.

[29] Larsen, C. 2003. Equality for the sexes in human evolution: Early hominid sexual dimorphism and implications for mating systems and social behavior. *Proceedings of the National Academy of Sciences* 100(16): 9103-9104.

[30] Reno, P. Meindl, R., McCollum, M, Lovejoy, C. 2003. Sexual dimorphism in *Australopithecus afarensis* was similar to that of modern humans. *Proceedings of the National Academy of Sciences,* 100(16): 9404-9409.

[31] Plavcan, J. Ward, C. & Paulus, F. 2009. Estimating canine tooth crown height in early Australopithecus. *Journal of Human Evolution* 57: 2-10.

[32] Lee, S. 2005. Patterns of size sexual dimorphism in *Australopithecus afarensis*: Another look. *Journal of Comparative Human Biology* 56: 219-232,

[33] Reno, P., Meindl, R., McCollum, M. & Lovejoy, C. 2005. The case is unchanged and remains robust: *Australopithecus afarensis* exhibits only moderate skeletal dimorphism. A reply to Plavcan et al. 2005. *Journal of Human Evolution* 49: 279-288.

[34] Crompton, Robin, et al. 2021. STW 573 Australopithecus Prometheus: Its Significance for an Australopith Bauplan. *Folia Primatology*, 92:243-275.

[35] Spoor, F, et al. 2016. Middle Pliocene hominin diversity: Australopithecus deyiremeda and Kenyanthropus platyops. *Philosophical Transactions B* 371: 20150231.

[36] Leakey et al. 2001. New hominin genus from eastern Africa shows diverse middle Pliocene lineages. *Nature* 410: 433-400.

[37] Lieberman, D. 2001. Another face in our family tree. *Nature,* 410:433-440.

[38] Cele-conde & Ayala, F. 2003. Genera of the human lineage. *Proceedings of the National Academy of Sciences* 100(3): 7684-7689.

[39] Harmand, S., & Lewis, J. (2023). “Lomekwi 3, West Turkana, Kenya.” In *Handbook of Pleistocene Archaeology of Africa: Hominin Behavior, Geography, and Chronology, Vol. 1*, edited by Amanuel Beyin et al. New York: Springer. p. 635.

[40] Harman, Sonia, 2015. 3.3-million-year-old stone tools from Lomekwi 3, West Turkana, Kenya. *Nature*, 521.

[41] Alemseged, Z. 2007. Paleoanthropological research at Dikika, Ethiopia, and comparative morphology of the juvenile hominin from DIK-1. *Paleonthropological Society Meeting Abstracts,* Philadelphia PA. 27-28 March 2007.

[42] Cameron, D. 2004. Hominid: Adaptations and Extinctions. (p. 180-182, 194). Sydney, Australia: University of New South Wales Press.

[43] Harmon, E. 2009. Size and shape variation in the proximal femor of Australopithecus africanus. *Journal of Human Evolution.*

[44] Green, D. & Gordon. 2006. Metacarpal proportions in Australopithecus Africanus. AAPA abstracts p. 119. *American Journal of Physical Anthropology.*

[45] Sponheimer, M. & Lee-Thorp, J. 2003. Differential resource utilization by extant great apes and *australopithecines*: Towards solving the C4 conundrum. *Comparative Biochemistry and Physiology, Part A* 136: 27-34.

[46] Campbell, B., Loy, J., Cruz-Uribe, K. 2005. *Humankind Emerging, 9th edition* (p. 344) Boston: Pearson, Allyn & Bacon.

[47] Holloway, R., Clark, R. Tobias, P. 2004. Posterior lunate sulcus in *Australopithecus africanus*: was Dart right? *Comptes Rendus Palevol,* 3(4): 285-292.

[48] See Quam R. et al. 2017. See section 8.51 Early Hominins. "Evolution of Hearing and Language in Fossil Hominins." *In Primate Hearing and Communication*, Springer. (& Quam, Dissertation p. 504.)

[49] Johnson, R. & Edgar, B. 2006. *From Lucy to Language* (p. 133) Simon and Schuster.

[50] Bednarik, R. 1993. The 'Australopithecine Cobble' from Makapansgat, South Africa, *South African Archaeological Bulletin* 53: 4-8.

[51] Bednarik, R. 2020. About the Origins of the Human Ability to Create Constructs of Reality. *Axiomatches*, Article number 9537.

[52] Semaw, S. et al. 2003. 2.6 - million-year-old stone tools and associated bones from OGS-6 and OGS-7, Gona, Afar, Ethiopia. *Journal of Human Evolution* 45:169-177.

[53] Delagnes, A. & Roche, H. 2005. Late Pliocene hominid skills: The case of Lokalalei 2C, West Turkana, Kenya. *Journal of Human Evolution* 48: 435-472.

[54] Arp, R. 2006. The environments of our Hominin ancestors, tool-usage, and scenario visualization. *Biology and Philosophy* 21(1): 95-117.

[55] Dominguez-Rodrigo, M., Pickering, T., Semaw, S. & Rogers, M. 2005. Cutmarked bones from Pliocene archaeological sites at Gona, Afar, Ethiopia: Implications for the function of the world's oldest stone tools. *Journal of Human Evolution* 48: 109-121.

[56] Rose, L. & Marshall, F. 1996. Meat eating, hominid sociality and home bases revisited. *Current Anthropology* 37: 307-338.

[57] King, B 2017. Evolving God: *A Provocative View on the Origin of Religion*, (p. 88 in earlier edition). University of Chicago Press.

[58] Johanson, D. & Edgar, B. 2006. *From Lucy to Language,* (p 131). New York: Simon and Schuster

[59] Roger Lewin & Robert Foley 2003. Principles of Human Evolution, 2nd edition. (p. 273) Blackwell Publishing.

[60] Glen Conroy & Herman Pontzer, 2012. *Reconstructing Human origins; A Modern Synthesis, Third Edition.* See section on *Australopithecus garhi*, p 249. W. W. Norton, New York.

[61] Delagnes, A. & Roche, H. 2005. Late Pliocene hominid skills: The case of Lokalalei 2C, West Turkana, Kenya. *Journal of Human Evolution* 48: 435-472.

[62] Kimbel, et al. 1996. Late Pliocene *Homo* and Oldowan tools from the Hadar formation, (Kada Hadar Member) Ethiopia. *Journal of Human Evolution* 31: 827-847,

[63] Stedman, H., Kozyak, B., Nelson, A. *et al.* 2004. Myosin gene mutation correlates with anatomical changes in the human lineage. *Nature* 428, 415–418. https://doi.org/10.1038/nature02358

[64] Weaver, A. 2005. Reciprocal evolution of the cerebellum and neocortex in fossil humans. *Proceedings of the National Academy of Sciences* 102(10): 3576-3580.

[65] Haeusler, M. & McHenry, H. 2004. Body proportions of *Homo habilis* reviewed. *Journal of Human Evolution* 46: 433-465.

[66] Hausler, M. & McHenry, H. 2007. Evolutionary reversals of limb proportions in early hominids? Evidence from KNM-ER 3735 (*Homo habilis*). *Journal of Human Evolution* 53: 383-405.

[67] Tuttle, R. 1967. Knuckle walking and the evolution of hominoid hands. *American Journal of Physical Anthropology* 26: 171-206.

[68] Barham, L. & Mitchell, P. 2008. *The First Africans.* (p. 112-113). Cambridge University Press.

[69] Karakostis, Fotios. 2021. Biomechanics of the human thumb and the evolution of dexterity. *Current Biology*, 31: 1317-1325.

[70] Mussi, Margherita, et al. 2023. Early Homo erectus lived at high altitudes and produce both Oldwoan and Acheulean tools. *Science* 382: 713-718.

[71] Baster, Marcus et al. 2020. Rib cage anatomy in Homo erectus suggests a recent evolutionary origin of modern human body shape. Nature ecology & evolution 4: 1178-1187.

[72] Steudel-Numbers, K. 2006. Energetics in *Homo erectus* and other early hominins: The consequences of increased lower-limb length. *Journal of Human Evolution* 51(5): 445-453.

[73] Bramble, D. & Lieberman, D. 2004. Endurance running and the evolution of *Homo*. *Nature* 432: 345-352.

[74] Barham, L. & Mitchell, P. 2008. *The First Africans.* (p. 146-148). Cambridge University Press.

[75] Mercader, Julio et al. 2021. Earliest Olduvai hominins exploited unstable environments ~ 2 million years ago. *Nature Communications* 12.3.

[76] Cobo-Sánchez, L., 2022. Tracking Early Hominin Foraging Behavior at DS Site (Bed I, Olduvai Gorge) Frühes menschliches Ernährungsverhalten: Beiträge aus der taphonomischen und räumlichen Untersuchung von DS (Bed I, Olduvai-Schlucht).

[77] Robert Boyd, Joan Silk & Kevin Langergraber 2024. For quote ee section on The Social Life of early *Homo erectus*... (p. 292.) In *How Humans Evolved 10th edition.* W.W. Norton & Company.

[78] Barr, Andrew W. 2022. No sustained increase in zooarchaeological evidence for carnivory after the appearance of Homo erectus. *PNAS*, 119(5).

[79] Cela-Conde, Camilo J., and Francisco J. Ayala. 2007. *Human Evolution: Trails from the Past.* (172–173.) Oxford University Press.

[80] Bromage, T.G., McMahon, J.M., Thackeray, J.F., Kullmer, O., Hogg, R., Rosenberger, A.L., Schrenk, F. & Enlow, D.H. 2008. "Craniofacial architectural constraints and their importance for

reconstructing the early Homo skull KNM-ER1470." *Journal of Clinical Pediatric Dentistry*, 33(1), 43–54.

[81] Holloway, R., Broadfield, D. Y Yuan M. 2004. Brain enclosures: The Paleoneurological Evidence, Vol 3. *The Human Fossil Record* JH Schwartz, I Tattersall (eds) New York: Wiley-Liss.

[82] Corballis, M. 2003. From mouth to hand: Gesture, speech, and the evolution of right-handedness. *Behavioral and Brain Sciences* 26: 199-260.

[83] Plummer, T. 2004. Flaked bones and old bones: Biological and cultural evolution at the dawn of technology. *Yearbook of Physical Anthropology,* 47: 118-164.

[84] Braun et al. 2008. Oldowan behavior and raw material transport: perspectives from the Kanjera Formation. *Journal of Archaeological Science*, 35: 2329-2345.

[85] Marwick, B. 2005. The interpersonal origins of language. *Linguistics and Human Science* 1(2): 197-224.

[86] Kingston, J., Ditchfield, P., Plummer, T., Bishop, L., Ferraro, J. 2005. Isotopic approaches to paleoecological reconstruction at Olduvai Bed I (Tanzania) and Kanjera (Kenya.). AAPA abstracts.

[87] Ungar, P., Grine, F., Teaford, M., Zaatari, S. 2006. Diet in early *Homo:* A review of the evidence and a new model of adaptative versatility. *Journal of Human Evolution* 50: 78-95.

[88] Dominguez-Rodrigo, M. & Barba, R. 2006. New estimates of tooth mark and percussion mark frequencies at the FLK Zinj site: The carnivore-hominid-carnivore hypothesis falsified. *Journal of Human Evolution* 50: 170-194.

[89] O'Connell, J. Hawkes, K. & Jones, N. 1988. Hazda scavenging: Implications for Plio/Pleistocene hominid subsistence. *Review of Anthropology* 29: 124-146.

[90] Lordkipanidze, D. et al. 2005. Anthropology: The earliest toothless hominin skull. *Nature* 434: 717-718.

[91] Meyer, M., Lordkipanidze, D. Vekua, A. 2005. Evidence for the anatomical capacity for spoken language in *Homo erectus*. AAPA abstracts, 74th Annual Meeting, Paleoanthropology Society, Milwaukee WI, April 5th and 6th.

[92] Toth, N. 1985. The Oldowan reconsidered. A close look at early stone artifacts. *Journal of Archaeological Science* 12: 101-120.

[93] Wenke, R. & Okszewski, D. 2006. *Patterns in Prehistory: Humankinds First Three Million Years.* (p.107). Oxford University Press.

[94] Harrod, J. 2007. Synopsis of Paleolithic Africa. *Originsnet web site.*

[95] Dominguez-Rodrigo, et al. 2023. Earliest Acheulian paleolandscape reveals a 1.7 million-year-old megasite at Olduvai Gorge (Tanzania). *Quaternary Science Reviews* 316.

[96] Ponce de Leon, Marica et al. 2021. The primitive brain of early Homo. *Science* 372: 165-171.

[97] Meyer, M. 2005. *Functional biology of the Homo erectus axial skeleton from Dmanisi, Georgia.* University of Pennsylvania, PhD dissertation. Ann Arbor, Michigan: Proquest Information and Learning Company.

[98] Anton, S. & Swisher, C. 2004. Early dispersals of *Homo* from Africa. *Annual Review of Anthropology,* 33: 271-296.

[99] Lewin, R. & Foley, R. 2004. *Principles of Human Evolution.* (p. 326) Oxford UK: Blackwell Publishing.

[100] Barham, L. & Mitchell, P. 2008. *The First Africans.* (p. 245). Cambridge University Press.

[101] Mithen, S. 2006. Ethnobiology and the evolution of the human mind. *Journal of the Royal Anthropological Institute* 12(s1): 45-61.

[102] De la Torre, Ignacio. 2016. The origins of the Acheulean: past and present perspectives on a major transitions in human evolution. *Philosophical Transactions R. Soc B*: 371.

[103] Langdon, John . 2022. Chapter 16: Behavior in the Middle Pleistocene (Section on Mode II Cultures) *Human Evolution: Bones, Cultures, and Genes.* Springer.

[104] Lepre, C. 2012. An earlier origin for the Acheulian. *Nature* 477: 82–85.

[105] Dominguez-Rodrigo, M., Serrallonga, J., Juan-Tresserras, J., Alcala, L., Luque, L. 2001. Woodworking activities by early humans: a plant residue analysis on Acheulian stone tools form Peninj (Tanzania). *Journal of Human Evolution* 40: 289-299.

[106] Gibbon et al. 2009. Early Acheulean technology in the Rietputs Formation, South Africa, dated with cosmogenic nuclides. *Journal of Human Evolution* 56:152-160.

[107] Personal communication, April 2005, with Bobby Dixon, flint knapper and collector of archaic Native American tools, Galax Virginia.

[108] Barham, L. & Mitchell, P. 2008. *The First Africans.* (p. 180). Cambridge University Press.

[109] Leary, M, & Buttermore, N. 2003. The evolution of the human self: Tracing the natural history of self-awareness. *Journal for the Theory of Social Behavior* 33(4): 365-404.

[110] Tensorer, J. 2005. Les cultures acheuleenes et la question de l'emergence de la pensee symbolique chez *Homo erectus* a partir des donnees relatives a la forme symetrique et harmonique des bifaces. *Comptes Rendus Palevol* 5(1,2): 127-135.

[111] Harrod, J. 1992. The way of the birth-giver: The religious imagination of the early and middle Acheulian (ca 600,000-300,000). *Paper presented to the American Academy of Religion Annual Meeting.* Chicago, November 19-22, 1988.

[112] Wynn, T. 2002. Archaeology and cognitive evolution. *Behavioral and Brain Research* 25: 389-438.

[113] Leary, M, & Buttermore, N. 2003. The evolution of the human self: Tracing the natural history of self-awareness. *Journal for the Theory of Social Behavior* 33(4): 365-404.

[114] Harod, James 2014. Palaeoart at Two Million Years? A Review of the Evidence. *Arts* 3: 135-155.

[115] Pickering, T. White, T. & Toth, N. 2000. Cutmarks on a Plio-Pleistocene hominid from Sterkfontein, South Africa. *American Journal of Physical Anthropology* 111: 579-584.

[116] Falk, D. 2004. Prelinguistic evolution in early hominins. Whence motherease? *Behavioral Brain and Sciences*, 27: 491-503.

[117] Hlubik, Sarah, et al. 2019. Hominin fire use in the Okote Member at Koobi Fora, Kenya: new evidence fo the old debate.

[118] Fitch, W. 2005. The evolution of language: A comparative review. *Biology and Philosophy* 20: 193-230.

[119] Corballis, M. 2002. *From Hand to Mouth: The Origins of Language,* Princeton: Princeton University Press.

[120] Masataka, N. 2009. The origins of language and the evolution of music: A comparative perspective. *Physics of Life Reviews.*

[121] Bickerton 2000. How protolanguage became language. In Knight, C. Studdert-Kennedy, M. And Hurford J. (Eds). *The Evolutionary Emergence of Language.* Cambridge University Press, Cambridge, Mass, pp. 264-284.

[122] Gibson, K. 1996. The ontogeny and evolution of the brain, cognition and language. In A. Lock & C. Peters.(eds) *A. Handbook of Human Symbolic Evolution.* Oxford: Clarendon Press.

[123] DeMenocal, P. 2004. African climate change and faunal evolution during the Pliocene-Pleistocene. *Earth and Planetary Science Letters* 6976: 1-22.

[124] Barham, L. & Mitchell, P. 2008. *The First Africans.* (p. 161). Cambridge University Press.

[125] Simpson SW, Quade J, Levin NE, Butler R, Dupont-Nivet G, Everett M, Semaw S. 2008. A female Homo erectus pelvis from Gona, Ethiopia. Science 322(5904):1089–1092. doi:10.1126/science.1163592

[126] Manzi, G. 2004. Human evolution at the Matuyama-Brunhes boundary. *Evolutionary Anthropology* 13: 11-24.

[127] Wang, W. & Crompton, R. 2003. Size and power required for motion with implication for the evolution of early hominids. *Journal of Biomechanics* 36: 1237-1246.

[128] Anton, S. 2003. Natural history of *Homo erectus. American Journal of Physical Anthropology* 122(S37): 126-170.

[129] Asfaw B. et al. 2002. Remains of Homo erectus from Bouri, Middle Awash, Ethiopia, *Nature,* 416, 317-320.

[130] Campbell, B., Loy, J., Cruz-Uribe, K. 2005. *Humankind Emerging, 9th edition* (p. 326). Boston: Pearson, Allyn & Bacon.

[131] Baab, K. 2006. Magnitude and pattern of geographic variation in cranial shape within *Homo erectus,* especially from Asia. AAPA abstracts, p. 66. *American Journal of Physical Anthropology.*

[132] See Feder's 2024 section on Chronicle (p. 111-112) in *The Past in Perspective: An Introduction to Human Prehistory*, Oxford University Press.

[133] Lordkipanidze, D. et al. 2013. A Complete Skull from Dmanisi, Georgia, and the evolutionary biology of modern *Homo*. *Science* 342: 326-331.

[134] Calvin, W. 2004. *A Brief History of the Mind; From Apes to Intellect and Beyond.* (p. 54). Oxford: Oxford University Press.

[135] Gowlett J. Harris, J., Walton D. & Wood B. 1981. Early archaeological sites, hominid remains and traces of fire from Chesowanja, Kenya. *Nature* 294: 125 - 129.

[136] Berna, F. et al 2012. Microstratigraphic evidence of in situ fire in the Acheulean strata of Wondewerk Cave, Northern Cape province, South Africa. *PNAS*, E1215-E1220.

[137] Berna, F. et al. 2012. Microstratigraphic evidence of in suit fire in the Acheulean strata Wonderwerk Cave, Northern Cape Province, South Africa. PNAS E1215-E1220.

[138] Ryan J. Gibbon, et al. 3014 Cosmogenic nuclide burial dating of hominin-bearing Pleistocene cave deposits at Swartkrans, South Africa, *Quaternary Geochronology* 24: 10-15.

[139] See section Control of Fire (p. 135) in Kenneth L. Feder's 2024 *The Past in Perspective: An Introduction to Human Prehistory*, Oxford University Press.

[140] Rincon, P. 2004. Early human marks are 'symbols,' (BBC report on find by Guadelli and others from the University of Bordeaux, March 16, 2004).

[141] Schick & Clark. 2003. Biface technological development and variability in the Acheulean Industrial Complex in the Middle Awash region of the Afar Rift, Ethiopia, in M. Sorresi and H.L Dibble (eds.) *Multiple Approaches to the Study of Bifacial Technologies*, Philadelphia: University Museum Monograph 115: 1-30.

[142] Hallos, J. 2005. "15 Minutes of Fame,": Exploring the temporal dimension of Middle Pleistocene lithic technology. *Journal of Human Evolution* 49: 155-179.

[143] Marwick, B. 2003. Pleistocene exchange networks as evidence for the evolution of language. *Cambridge Archaeological Journal* 13: 67-81.

[144] Morwood, M. , O'Sullivan P., Aziz F. Raza A. 1998. Fission-track ages of stone tools and fossils on the east an island of Flores *Nature* 392: 173-176.

[145] Manzi, G. 2004. Human evolution at the Matuyama-Brunhes boundary. *Evolutionary Anthropology* 13:11-24.

[146] Abbate, et al. 1998. A one million year old *Homo* cranium from the Danakil (Atar) depression of Eritea. *Nature* 393: 458-460.

[147] Berger, G. Et al. 2008. Luminescence chronology of cave sediments at the Atapuerca paleoanthropological site, Spain. *Journal of Human Evolution* :1-12

[148] De Castro, J., Martinon-Torres, M., Carbonell, E., Sarmiento, S., Rosas, A.Van der Made, J., Lozano, M. 2004. The Atapuerca sites and their contribution to the knowledge of human evolution in Europe. *Evolutionary Anthropology* 13: 25-41.

[149] Arsuaga, J. & Martinez, I. 2006. *The Chosen Species,* (English translation of 1998 edition) p. 188. New York: Blackwell Publishing.

[150] de Castro et al. 1999. A modern human pattern of dental development in Lower Paleolithic hominds from Atapuerca-TD6 (Spain). *PNAS* 96 (7).

[151] Zilhao, J. 2006. Neandertals and moderns mixed, and it matters. *Evolutionary Anthropology,* 15: 183-196.

[152] Quoted in El mundo. 9/19/06 El '*Homo Antecessor*' de Atapuerca era diestro.

[153] Fernandez-Jalvo. Y., Diez, J. Caceres, I., Rosell, J. 1999. Human cannibalism in the Early Pleistocene of Europe (Gran Dolina, Sierra de Atapuerca, Burgos, Spain. *Journal of Human Evolution,* 37:591-622

[154] Petrinovich, L. 2000. *The Cannibal Within* (Evolutionary Foundations of Human Behavior). New York: Aldine Transactions.

[155] American Museum of Natural History, 2004. Web site: "The Humans of Gran Dolina." www..amnh.org/exhibitions/atapuerca/granhumans.

[156] Bruner, E & Manzi, G. 2007. Landmark-based shape analysis of the archaic *Homo* calvarfium from Ceprano, Italy. *American Journal of Physical Anthropology,* 132: 355-366.

[157] Vialet et al. 2004. Proposition de reconstitution du deuxième crâne d'Homo erectus de Yunxian (Chine). *Comptes Rendus Palevol* 4(3): 265-274.

[158] Campbell, B., Loy, J., Cruz-Uribe, K. 2005. *Humankind Emerging, 9th edition* (see p. 345) Boston, Pearson, Allyn & Bacon.

[159] Goren-Inbar, N. Lister, A. Weker, E. Chech, M. 1994. A butchered elephant skull and associated artifacts from the Acheulian site of Gesher Benot Ya'Aqov, Israel. *Paleorient* 20(1): 99-112.

[160] Frisson, G. 2004. *Survival by hunting: Prehistoric Human Predators and Animal Prey.* Berkeley: University of California.

[161] Goren-Inbar, N., Sharon, G., Melamed, Y, Kisleve, M. 2002. Nuts, nut cracking, and pitted stones at Gesher Benot Ya`aqov, Israel. Proceedings of the National Academy of Sciences 99(4): 2455-2460.

[162] Barham, L. & Mitchell, P. 2008. *The First Africans.* (p. 144). Cambridge University Press.

[163] Alperson-Afil, Nira. 2017. Spatial Analysis of Fire: Archaeological Approach to Recognizing Early Fire. *Current Anthropology*, 58, S16.

[164] Alperson-Afil, Richter, Daniel, & Goren-Inbar, Naama. 2017. Evaluating the intensity of fire at the Acheulian site of Geshser Benot Ya'aqov—Spatial and thermoluminescence analysis. *Plos One* 12(11).

[165] Alperson-Afil, N. 2008. Continual fire-making by Hominins at Gesher Benot Ya'aqov, Israel. *Quaternity Science Review* 27: 1733-1739.

[166] Goren-Inbar, N. & Saragusti, I. 1996. An Acheulian biface assemblage from Gesher Benot Ya'aqov, Israel: Indications of African affinities. *Journal of Field Archeology* 23 (1): 15-30.

[167] Personal communication with Gonen Sharon of The Institute of Archeology, The Hebrew University of Jerusalem, e-mail February 25, 2007.

[168] Goren-Inbar et al. 2018. *The Acheulian Site of Gesher Benot Ya'aqov Volume IV: The Lithic Assemblages*. NY: Springer.

[169] Goren-Inbar et al. 2000. Pleistocene milestones on the out-of-Africa corridor at Gesher Benot Ya'aqov, Israel. Science. 289 (5481): 944-947.

[170] Personal communication with Gonen Sharon of The Institute of Archeology, The Hebrew University of Jerusalem, e-mail February 25, 2007.

[171] Beaumont P. & Vogel, J. 2006. On a timescale for the past million years of human history in central South Africa. *South African Journal of Science* 102: 217-228.

[172] Calvin, W. 2004. *A Brief History of the Mind; From Apes to Intellect and Beyond*. (p. 54). Oxford: Oxford University Press.

[173] Wynn, T. 2002. Archaeology and cognitive evolution. *Behavioral and Brain Research* 25: 389-438.

[174] Shipton, Ceri, 2022. Predetermined Refinement: The Earliest Levallois of the Kapthurin Formation. *Journal of Paleolithic Archaeology*, 5:4.

[175] Rosenberg-Yefet, et al. 2022. Lower Paleolithic Winds of Change: Prepared Core Technologies and the Onset of the Levallois Method in the Levantine Late Acheulian. *Frontiers in Earth Science.*

[176] See section on Prepared Cores (p. 1830) from the important review by Deborah I. Olszewki et al. "The Pleistocene Stone Artifact Record of Africa: Technologies, Typologies, and Analytic Approaches" *in Handbook of Plesistocene Archaeology of Africa: Hominin Behavior, Geography, and Chronology, Volume 2;* edited by Amanuel Beyin et al. 2023 NY: Springer.

[177] Goren-Inbar et al. 2018. See section 9.3 Final Words (p. 424) *The Acheulian Site of Gesher Benot Ya'aqov Volume IV: The Lithic Assemblages*. NY: Springer.

[178] Di Vincenzo, F., & Manzi, G. 2023. Homo heidelbergensis as the Middle Pleistocene common ancestor of Denisovans, Neanderthals and modern humans. *Journal of Mediterranean Earth Sciences, 15.* https://doi.org/10.13133/2280-6148/18074.

[179] See Kathleen Kuman's section on Acheulian Industrial Complex (p. 14) in *Encyclopedia of Global Archaeology*, Claire Smith (editor), Spring Science-Buisness Media.

[180] Klein, R. 2005. Hominin dispersals in the old world. In *The Human Past,* ed. Scarre, C. (p. 107) London: Thames and Hudson.

[181] Rightmire G. 2003. Brain size and encelphalization in early to Mid-Pleistocene *Homo, American Journal of Physical Anthropology* 124 (2): 109-123.

[182] Barham, L. & Mitchell, P. 2008. *The First Africans.* (p. 211). Cambridge University Press.

[183] Maclarnon, A. & Hewitt, G. 2004. Increased breathing control: Another factor in the evolution of human language. *Evolutionary Anthropology* 13:181-197.

[184] Jeusel, E. & Marfat, B. 2007. Morphological evolution of the bony palate of hominid fossils. *Paleonthropological Society Meeting Abstracts,* Philadelphia PA 27-28 March 2007.

[185] Dunbar, R. 2004. *The Human Story*. London: Faber & Faber.

[186] Quam, R. 2006. Temporal bone anatomy and the evolution of acoustic capacities in fossil humans. (p. 5050) Dissertation, Submitted for Doctorate of Philosophy in Anthropology in the Graduate School of Binghamton University, State University of New York.

[187] Martinez, et al. 2007. Two Middle Pleistocene human hyoid bones from the Sima de los Huescos site (Sierra de Atapuerca, Spain) *Paleonthropological Society Meeting Abstracts,* Philadelphia PA 27-28 March 2007.

[188] Martinez, I. Rosa, M. Arsuaga, J., Jarabo, P., Quam, R. Lorenzo, C. Gracia, A. Carretero. J. de Castro, J., Carbonell, E. 2004. Auditory capacities in Middle Pleistocene humans from the Sierra de Atapuerca in Spain. *Proceedings of the National Academy of Sciences* 101(27): 9976 -9981.

[189] Urquiza, R., Botella, M. & Ciges, M. 2005. Study of a temporal bone of *Homo heidelbergensis. Acta Oto-Laryngologica* 125: 457-463.

[190] Millard, A. 2007. A critique of the chronometric evidence for hominid fossils I. Africa and the Near East 500-50 ka. *Journal of Human Evolution.*

[191] White, T. 2005. Cut marks on the Bodo cranium: A case of prehistoric defleshing. *American Journal of Physical Anthropology* 69(4): 503-509.

[192] Hayden, Brian 2003. *Shamans, Sorcerers and Saints*: *A Prehistory of Religions* (p. 97) Smithsonian Books.

[193] Muttoni, Giovanni, et al. 2017. An early Brunhes (<0.78 Ma) age for the Lower Paleolithic tool-bearing Kozarnika cave sediments, Bulgaria. (A careful reading of the article reveals the dating of Layer 12 as greater than 500 Kya with an outer limit of 600 Kya.) *Quaternary Science Reviews* 178: 1-13.

[194] Sirakov, N. Guadelli J.-L, et al 2010. An ancient continuous human presence in the Balkans and the beginnings of human settlement in western Eurasia. A lower Pleistocene example of the Lower Palaeolithic levels in Kozarnika cave (North-western Bulgaria). (See Fig. 14 for photo of shin bone & 6. Conclusion for quote.) *Quaternary International* 223-224: 94-106.

[195] See the important article by Ian Watts, Michael Chazan and Jayne Wilson. Early evidence for Brilliant Ritualized display: Specularite Use in the Northern Cape (South Africa) between 500 and 300 Ka. *Current Anthropology* 57:3. 2016.

[196] See section on The Earliest Evidence for Ochre Use in "Complex Colors: Pleistocene Ochre Use in Africa" by Tammy Hodgskiss in *Handbook of Plesistocene Archaeology of Africa: Hominin Behavior, Geography, and Chronology, Volume 2.* 2023 NY: Springer.

[197] Bednarik, R. 1994. Art origins. *Anthropos* 89: 169-180

[198] See section on Archeological Sequence from Wonderwerk Cave and the Kathu Compelx, South Africa by Michael Chazan & Liora Horwitz from *Handbook of Plesistocene Archaeology of Africa: Hominin Behavior, Geography, and Chronology, Volume 2,* edited by Amanuel Beyin et al. 2023 NY: Springer.

[199] See the above cite, section on The Back of the Cave.

[200] Stafford, M., Frison, G., Stanford, D. & Zeimans, G. 2003. Digging for the color of life: Paleoindian red ochre mining at the powars II site, Platte County, Wyoming, U.S.A. *Geoarchaeology: An International journal* 18(1): 71-90.

[201] Knight, C. Power C., & Watts, I. 2003. In comments on "An early case of color symbolism." *Current Anthropology* 44(4): 513-514.

[202] Hayden, Brian 2003. *Shamans, Sorcerers and Saints: A Prehistory of Religions* (p. 99) Smithsonian Books.

[203] See Theodora Mousiou's "Colour in the Paleolithic," in the *Value of Colour: Material and Economic Aspects in the Ancient World.* 2019. Edited by Shiyanthi Thavapalan & David Alan Warburton. Berlin Studies of the Ancient World.

[204] Hogenson, G. 2004. What are symbols symbols of? Situated action, mythological bootstrapping and the emergence of the Self. *Journal of Analytic Psychology* 49: 67-81.

[205] Personal communication by e-mail; March 30, 2004.

[206] Wilkens, J. et al. 2012 Evidence for Early Hafted Hunting Technology. *Science* 338: 942-946.

[207] Deino, A., & McBrearty, S. 2002. ^{40}Ar/^{39}Ar dating of the Kapthurin formation, Baringo, Kenya. *Journal of Human Evolution* 42: 185-210.

[208] Johnson, C & MCreaty, S. 2010. 500,000 year old blades from the Kapthurin Formation, Kenya. *Journal of Human Evolution* 58: 193-200.

[209] Blegen, Nick, Brian Jicha & Sally McBrearty 2018. A New Tephrochronology for Early Diverse Tool Technologies and Long-Distance Raw Material Transport in the Middle to Late

Pleistocene Kapthurin Formation, East Africa. *Journal of Human Evolution* 121:75-103..

[210] See Kenneth L. Feder's 2024 very lucid and helpful work, his section on Stone Tools of Anatomically modern Human beings : Utilitarian works of art. (p 173-175) in *The Past in Perspective: An Introduction to Human Prehistory*, Oxford University Press.

[211] Barham, L. et al. 2023 Evidence for the earliest structural use of wood at least 476 Kya. *Nature* 622: 107- 122.

[212] Meyer, M., Arsuaga, JL., de Filippo, C. *et al.* Nuclear DNA sequences from the Middle Pleistocene Sima de los Huesos hominins. *Nature* **531**, 504–507 (2016).

[213] See Robert Boyd, Joan Silk & Kevnin E. Langergraber's good summation of Neanderthals from the Section on Neanderthals (pp. 303-313) in in *How Humans Evolved 10th edition.* W.W. Norton & Company, 2024.

[214] Arnold, Lee, et al. 2014. Luminescence dating and palaeomagnetic age constraint on hominins from Sima de los Huesos, Atapuerca Spain. *Journal of Human Evolution* 67: 85-197.

[215] See section on The Origin of Accumulation/El Origen de la Acumulacion (p 16). Mosquera, M. Atapuerca: El gran tesora arqueologico que ilumina los secretos de la evolucion humana. 2024. Pinola.

[216] de Castro, J., Martinon-Torres, M., Sarmiento, S., and Lozano, M. 2003. Gran Dolina-TD6 versus Sima de los Huescos dental samples from Atapuerca: evidence for discontinuity in the European Pleistocene population. *Journal of Archaeological Science* 30(11): 1421-1428.

[217] Carbonell, E., & Mosquera, M. 2006. The emergence of symbolic behavior: The sepulchral pit of Sima de los Huescos, Sierra de Atapuerca, Burgos, Spain. *Comptes Rendus Palevol* 5(1,2): 155-160.

[218] King, B. 2007. Evolving God: A provocative view on the Origins of Religon. New York: Doubleday.

[219] Johnson, R. & Edgar, B. 2006. *From Lucy to Language.* (p. 96) Simon and Schuster.

[220] Gowlett, J. 2006. The early settlement of northern Europe: Fire history in the context of climate change and the social brain. *Comptes Rendus Palevol* 5: 299-310.

[221] Lumley, H. 2006. Il y 400 000 ans: la domestication du feu, un formidable moteur d' hominization. *Comptes Revue Palevol* 5 (1,2): 149-154.

[222] Thieme, H. 2005. The Lower Paleolithic art of hunting: The case of Schoningen. In *The Hominid Individual in Context.* C. Gamble and M. Porr, Eds. Pp. 115-132. Ondon: Routledge.

[223] Stiner, M. 2002. Carnivory, coevolution, and the geographic spread of the Genus *Homo. Journal of Archaeological Research* 10(1): 1-62.

[224] Laidler, P. 1933. Dating evidence concerning the Middle Stone Ages and a Capsio-Wilton culture, in the South-East Cape. *South African Journal of Science,* 30: 530-542.

[225] Davis, W. 1993. Beginning the history of art, *The Journal of Aesthetics and Art Criticism* 51(3): 327-350.

[226] Bednarik, R. 1995. Concept-mediated marking in the lower Paleolithic. *Current Anthropology* 36(4): 605-634.

[227] Bednarik, R. 2003. A figurine from the African Acheulian. *Current Anthropology* 44(1) 405-413.

[228] Bahn, P. 2007. The earliest imagery around the globe. In C. Renfrew & I. Morley (editors) *Image and Imagination.* Cambridge: McDonald Institute for Archaeological Research.

[229] Barham, L. & Mitchell, P. (2008). *The First Africans.* (p. 205). Cambridge University Press.

[230] Grün R, Pike A, McDermott F, Eggins S, Mortimer G, Aubert M, Kinsley L, Joannes-Boyau R, Rumsey M, Denys C, Brink J, Clark T, Stringer C. 2020 Dating the skull from Broken Hill, Zambia, and its position in human evolution. *Nature*. 580(7803):372-375. doi: 10.1038/s41586-020-2165-4. Epub 2020 Apr 1. PMID: 32296179.

[231] Rightmire, G.P. 2017. Middle Pleistocene *Homo* Crania from Broken Hill and Petralona: Morphology, Metric Comparisons, and Evolutionary Relationships. In: Marom, A., Hovers, E. (eds) *Human Paleontology and Prehistory. Vertebrate Paleobiology and Paleoanthropology*. Springer, Cham. https://doi.org/10.1007/978-3-319-46646-0_1,1

[232] See chapter on "Jeberl Irhoud (Ighoud), Morocco" by Ben-Ncer A. et al (p 803-811) *in Handbook of Plesistocene Archaeology of Africa: Hominin Behavior, Geography, and Chronology, Volume 1.* edited by Amanuel Beyin et al. 2023 NY: Springer

[233] Bruner, Emiliano, and Osbjorn Pearson. 2013. Neurocranial evolution in modern humans: the case of Jebel Irhoud 1. *Anthropological Science* 121.1: 31-41.

[234] Robert Boyd, Joan Silk & Kevin Langergraber. 2024. See section on The earliest anatomical modern human fossils come from Africa and are dated between 315 and 160 ka. (pp 327-328). In *How Humans Evolved 10th edition.* W.W. Norton & Company.

[235] Neubauer, S. Hublin, J. & Gunze, P. 2018. The evolution of the modern human brain shape. *Science Advances* 4.

[236] Ben-Neer, A. et al. 2023. Chapter on Jebel Irhoud (Ighoud), Morocco See section on *The Hominin Fossil Remains* (p. 806-807) In *Handbook of Pleistocene Archaeology of Africa: Hominin Behavior,*

Geography, and Chronology, Volume 1, edited by Amanuel Beyin et al. Springer.

[237] Gunter Brauer, Yuji Yokoyama, Christophe Falguèeres, Emma Mbua. Modern human origins backdated. news@nature.com, 1997. ffhal-03741311f.

[238] Brink, James, 1997. Direct Dating of the Florisbad Fossil Hominid CULNA, September, 1997.

[239] Berger, L. R., & Hawks, J. 2023. Revisiting the age of the Florisbad hominin material. Homo – Journal of Comparative Human Biology, 74(1), 55–60.

[240] Bruner, E., & Lombard, M. 2020. The skull from Florisbad: A paleoneurological report. *Journal of Anthropological Sciences*, 98, 89–97.

[241] Rosenberg, K. Zune, L.& Buff, C. (2000). Body size, body proportions, and encelphalization in a Middle Pleistocene archaic human from northern China. *Proceedings of the National Academy of Sciences* 103(10): 3552-3556.

[242] Liu, Wu, et al. 2022 Hominin evolution and diversity: a comparison of earlier-Middle and later-Middle Pleistocene hominin fossil variation in China. *Philosophical Transactions of the Royal Society B* 377.1847: 20210040.

[243] Feder, K. 2024. See sections on A new Hominin Star and Where Does Homo Naledi fit in the story of Human Evolution? (pp. 123-126) In *The Past in Perspective: An Introduction to Human Prehistory, Ninth Edition.* Oxford University Press.

[244] Moncel, M., Moigne, A. & Combier, J. 2005. Pre-Neandertal behavior during isotopic stage 9 and the beginning of stage 8. *Journal of Archaeological Science* 32: 1283-1301.

[245] Tyron, C. 2006. "Early" Middle stone Age lithic technology of the Kapthurin Formation (Kenya). *Current Anthropology* 47(2): 367-374.

[246] Lewin, R. & Foley, R. 2004. *Principles of Human Evolution.* (p. 424) Oxford UK: Blackwell Publishing.

[247] Barham, L. 2003. Systematic pigment use in the Middle Pleistocene of South-Central Africa. *Current Anthropology* 43(1): 181-190.

[248] Power, C. & Watts, I. 1997. Female proto-symbolic strategies. In *Women in Evolution,* ed. L. Hager. P. 153-171. London: Routlege.

[249] D'Errico, F. & Norwell, A. 2000. A new look at the Berekhat Ram Figurine: Implications for the origin of symbolism. *Cambridge Archaeological Journal* 10(1): 123-167.

[250] *Highlights of the Archaeology, Art, Judaica and Jewish Ethnography collections,* 2005. The Israel Museum, Jerusalem http://www.imj.org.il/imagine/HightLight.asp

[251] See Tobi Zusner's Sectoin on Lower Paleolithic Figurines: The Bereket Ram and the Tan-Tan. From Chapter 5: "Evolution, Altered States and Creative Trance" In *The Creative Trance: Altered States of Consciousness and the Creative Process.* 2022. Cambridge University Press.

[252] Susko, M. 2021. *The Generation of Life. Imagery, Rituals, and Experiences in Deep Caves.* AllrOneofUs Publishing.

[253] Lorente, Aracell, 2020. The Origin of Art: An Approximation through Archaeological Evidences. *Sociology and Anthropology* 8(3): 82-91.

[254] Bednarik, R. 2001. Beads and Pendants of the Pleistocene. (see figure 1 with commentary on p. 549) *Anthropos* 96: 545-555.

[255] Barham, L. & Mitchell, P. 2008. *The First Africans.* (p. 219-221). Cambridge University Press.

[256] Pennisi, E. 1999. Did cooked tubers spur the evolution of big brains? *Science* 283: 5410(26): 2004-2005.

[257] Bednarik, R. 2001. The oldest known rock art in the world. *Anthropologie* 39(2-3): 89-97.

[258] Personal communication, Robert Bednarik, e-mail dated April 30, 2008.

[259] Bednarik R. 2008. Cupules. *Rock Art Research* 25(1): 61-100.

[260] Flood, J. 2006. Copying the Dreamtime: Anthropomorphic marks in early Aboriginal Australia. *Rock Art Research*, 23(1).

[261] Rots, V. & Van Peer, P. 2005. Early evidence of complexity in lithic economy: Core-axe production, hafting and use at Late Middle Pleistocene site 9-B-11, Sai Island (Sudan). *Journal of Archeological Science* 33(3): 360-371.

[262] Bednarik, R. 1997. About Ostrich Eggshell Beads. (See Figure 1 and commentary on p,156) Acta Archaeologica 68: 153-161.

[263] Bednarik, R. 2015. The Significance of the Earliest Beads. *Advances in Anthropology* 5: 51-66.

[264] Wynn, T. & Coolidge, F. 2004. The expert Neandertal mind. *Journal of Human Evolution* 46: 467-487.

[265] Calvin, W. 2005 *A Brief History of Mind. From Apes to Intellect and Beyond.* (pp. 90-930) Oxford University Press.

[266] Neubauer, S. Hublin, J, & Gunz, P. 2018. The evolution of modern human brain shape. (See Table I and Figure 2). *Science Advances*, 4: eaai5961.

[267] Ingman, M, Gyllensten, U. 2001. Analysis of the Complete Human mt6DNA Genome: Methodology and Inferences for human evolution, *Genome Analysis* 92: 454-461.

[268] Relethford, J. 2005. Studies of genetic variation and their relevance to questions and the origins and evolution of the human species. *AAPA Abstracts.*

[269] Trinkaus, E. 2005. Early Modern Humans. *Annual Review Anthropology*, 34: 207-30.

[270] McDougall, I., Brown, F., Fleagle, J. 2005. Stratigraphic placement and age of modern humans from Kibish, Ethiopia. *Nature* 433: 733-736

[271] Vidal, Céline M., et al. 2022. "Age of the oldest known Homo sapiens from eastern Africa." *Nature* 601.7894: 579-583.

[272] Neubauer, Simon, Jean-Jacques Hublin, and Philipp Gunz. 2018. "The evolution of modern human brain shape." (see section on Discussion, *The Brain of H. sapiens*, p. 3) *Science advances* 4.1.

[273] Smith, T. Et al. 2007. Earliest evidence of human life history in North African early *Homo sapiens. Proceedings of the National Academy of Sciences* 104(5): 6128-6133.

[274] Voight B. Et al 2005. Interrogating multiple aspects of variation in a full resequencing data set to infer human population size changes. *Proceedings of the National Academy of Sciences* 102: 18508-18513.

[275] White, T. Asfaw, B., DeGuasta, D., Gilbert, H., Richards, G. Suwa, G., Howell, F. 2003. Pleistocene *Homo* remains from Middle Awash, Ethiopia. *Nature* 423: 742-747.

[276] Clark, J., Beyene, Y., WoldeGabriel, G. Hart, W., Renne, P. Defleur, A., Suwa, G, Katoh, S. Ludwig, K., Boisserie J. Asfaw, B., White, T. 2003. Stratigraphic, chronological and behavioral contexts of Pleistocene *Homo sapiens* from Middle Awash, Ethiopia. *Nature* 423 (6967): 747-52

[277] Fagan, B. 2007. *People of the Earth: Introduction to World Prehistory, 12th edition*. (p. 10.) Upper Saddle River, New Jersey: Pearson, Prentice Hall.

[278] Zaleski, P. & Zaleski, C. 2005. *Prayer: A History*. (p. 10). Boston: Houghton Mifflin Company

[279] Marwick, B. 2003. Pleistocene exchange networks as evidence for the evolution of language. *Cambridge Archaeological Journal* 13: 67-81.

[280] Sehasseh, El Mehdl et al. 2021 Early Middle Stone Age personal ornaments from Bizmoune Cave, Essaouira, Morroco. *Science Advances* 7: eabi8620.

[281] Bednarik , R. 2006. The Middle Paleolithic Engravings from Oldisleben, Germany Anthropologie, 44 (2): 113-121.

[282] Trinkaus, E. 2005. Early Modern Humans. *Annual Review in Anthropology* 34: 207-30.

[283] Vandermeersch, B. 2005. Ce que nous apprennent les premieres sepultures. *Comptes Rendus Palevol* 5(1,2): 161-167.

[284] Mellars, P. 2005. The Impossible Coincidence: A Single Species Model For the Origins of Modern Human Behavior in Europe. (p.17 for quote) *Evolutionary Anthropology* 14:12-27.

[285] Barham, L. & Mitchell, P. 2008. *The First Africans.* Cambridge University Press.

[286] Vanhaeran et al. 2006. Middle Paleolithic shell beads in Israel and Algeria. *Science* 312, 1785-8.

[287] Hovers, E., Vandermeersch & Bar-Yosef, B. 1997. A Middle Paleolithic engraved artefact from Qafzeh Cave, Israel. *Rock Art Research* 14(2): 79-87.

[288] Hovers, E, Ilani, S., Bar-Yosef, O.,Vandermeersch, B. 2003. An early case of color symbolism: Ochre use by modern humans in Qafzeh Cave. *Current Anthropology* 44(4): 491-522.

[289] Patricia Rice & Norah Moloney 2008. *Biological Anthropology and Prehistory: Exploring Our Human Ancestory.* 2nd edition. (p. 187) Routledge.

[290] Wolpoff, M. Et al. 2004. Why not the Neandertals? *World Archaeology* 36(4): 527-546.

[291] Kelley Harris, Rasmus Nielsen, 2016. The Genetic Cost of Neanderthal Introgression, *Genetics* 203 (2): 881–891

[292] Andra Meneganzin, Massimo Bernardi, 2023. Were Neanderthals and Homo sapiens 'good species'? *Quaternary Science Reviews* 303.

[293] Mellars, P. 2005. The impossible coincidence. A single-species model for the origins of modern human behavior in Europe. *Evolutionary Anthropology* 14: 12-27.

[294] Yellen, et al. 1995. A Middle Stone Age worked bone industry from Katanda, upper Semliki Valley, Zaire. *Science* 268: 553-556.

[295] Soressi, M. & Henshilwood, C. 2004. Blombos cave, South Africa: Stone artifacts from the Middle Stone Age Levels, *Paleoanthropological Society Annual Meeting,* Montreal Canada.

[296] Jacobs, Z., Duller, G., Wintle, A., Henshilwood, C. 2006. Extending the chronology of deposits at Blombos Cave, South Africa, back to 140 ka using optical dating of single and multiple grains of quartz. *Journal of Human Evolution* 51: 255-273.

[297] Henshilwood, C., d'Errico, F. Marean, C., Milo, R. Yates, R. 2001. An early bone tool industry from the Middle Stone Age at Blombos Cave, South Africa: implications for the origins of modern

human behavior, symbolism and language. *Journal of Human Evolution* 41: 631-678.

[298] d' Errico, Hensilwood, C., Lawson, G., Vanhaeren, M., Tillier, A., Soressi, M., Bresson, F., Maureille, B., Nowell, A., Lakarra, J., Blackwell, L, & Julien, M. 2003. Archaeological evidence for the emergence of language, symbolism, and music – An alternative multidisciplinary perspective. *Journal of World Prehistory* 17(1): 1-70.

[299] Grun, R. 2006. Direct dating of human fossils. *Yearbook of Physical Anthropology* 49: 2-48.

[300] Thackeray, A, 1992. The Middle Stone Age south of the Limpopo River. *Journal of World Prehistory* 6: 385-440.

[301] Movius, H. L., 1953. The Mousterian cave of Teshik-Tash, Southeastern Uzbekistan, Central Asia. *Bulletin of the American School of Prehistoric Research*, 17, 11-71.

[302] Hayden, B. 2003. *Shamans, Sorcerers and Saints: A Prehistory of Religion.* (See section on Animals Dominated Early Pantheon, pp. 105-108) Washington: Smithsonian books.

[303] See section on Neanderthal made cave art (pp. 311-312) in Robert Boyd, Joan Silk & Kevin Langergraber. in *How Humans Evolved 10th edition.* W.W. Norton & Company, 2024.

[304] Bednarik, R. 2003. The earliest evidence of Palaeoart. *Rock Art Research*, 20 (2): 89-135.

[305] Leroi-Gourhan, A. 2000. Rites et Langage a Shanidar? *Bulletin de la Societe Prehistorique Francaise* 97(2): 291-293.

[306] Pomeroy, Emma, et al. 2020. New Neanderthal remains associated with the "flower burial" at Shanidar Cave. *Antiquity* 94(273): 11-26.

[307] Pettitt P. 2005. The rise of modern humans. (Chapter 4) In *The Human Past,* ed. Scarre, C. London: Thames and Hudson.

[308] Parkington, et al. 2005. From tool to symbol: the behavioral context of intentionally marked ostrich eggshell from Diepkloof, Eastern Cape. In (eds) D'Errico, F. & Blackwell, L. *From Tools to Symbol: From Early Hominins to Modern Humans.* Johannesburg: Witwatersrand University Press, p 475-492.

[309] Marshack, A. 1996. A Middle Paleolithic Symbolic Composition from the Golan Heights: The earliest known depictive image. *Current Anthropology* 37 (2): 357-365.

[310] Oktaviana, Adhi, et al. 2024. Narrative cave art in Indonesia by 51,200 years ago. *Nature* 631: 814-821.

[311] Hayden, B. 2003. *Shamans, Sorcerers and Saints: A Prehistory of Religion.* (p. 102-103). Washington: Smithsonian books.

[312] Nobbs, M. & Dorn, R. 1993. New surface exposure ages for petroglyphs from the Olary province, South Australia. *Archaeology in Oceania* 28: 18-39.

[313] Trinkaus, E. 2005. Early Modern Humans. *Annual Review in Anthropology* 34:207-30.

[314] Plumet, P. 2006. The far North and Religion. *L'anthropologie* 110: 383-400.

[315] Mellars, P. 2006. Archeology and the dispersal of modern humans in Europe: Deconstructing the "Aurignacian". *Evolutionary Anthropology*, 15: 167-182.

[316] Zilhao, J. & d'errico, F. 1999. The chronology and taphonomy of the earliest Aurignacian and its implications for the understanding of Neandertal Extinction. *Journal of World Prehistory* 13(1).

[317] Conrad, N. Bolus, M. 2003. Radiocarbon dating the appearance of modern humans and timing of cultural innovations in Europe: New results and new challenges. *Journal of Human Evolution* 44: 331-337.

[318] Fitch, W. 2006. The biology and evolution of music. *Cognition* 100: 173-215.

[319] Conard, N. 2009. A female figurine from the basal Aurignacian deposits of Hohle Fels Cave in southwestern Germany. *Nature* 459 (7244): 248

[320] Kvavadze, E., Bar-Yosef, O., Belfer-Cohen, A., Boaretto, E., Jakeli, N., Matskevich, Z., & Meshveliani, T. 2009. 30,000-year-old wild flax fibers. *Science*, 325(5946), 1359.

[321] Conrad, N. 2003. Paleolithic ivory sculptures from southwestern Germany and the origins of figurative art. *Nature* 426: 830- 832.

[322] See English Summary, The Data (pp. 127-132) and the beautiful illustrations in the Catalog (p 372-444) in Ewa Dutkiewicz, *Zeichen: Markierungen, Muster, und Symbole im Schwabischen Aurignacien*. 2021. Kerns Verlag Tubingen.

[323] Marshack A. 1972. Upper Paleolithic notation and symbol. *Science* 178: 817-828.

[324] Hoffecker, J. 2005. Innovation and technological knowledge in the Upper Paleolithic of Northern Eurasia. *Evolutionary Anthropology* 14: 186-198.

[325] Chauvet, Jean Marie, et al. 1996 Dawn of Art: *The Chauvet Cave: The Oldest Known Paintings in the World*. Harry N. Abrams.

[326] Quiles, Anita, et al. 2016. A high-precision chronological model for the decorated Upper Paleolithic cave of Chauvet-Pont d'Arc, Ardeche, France *PNAS* 113(17).

[327] See Conclusion: Shared Myths? In Carole Fritz and Gilles Tosellos's 2015, "From Gesture to Myth: L Artists' techniques on the walls of Chauvet Cave, *Palethnologie* 7.

[328] Phillipe M & Fose P.2003 Fauna of the Chauvet Cave (Vall-Pont-d'Arc Ardeche): preliminary paleontological and taphonomic presentation. *Paleo* 15.(Google Translation.)

[329] Goebel, T. 2004. The early Upper Paleolithic of Siberia. In (ed). Brantingham, J., Khuhn, S & Kerry, K. (p. 180.) The Early Upper Paleolithic Beyond Western Europe. Berkeley: University of California Press.

[330] Jacobi, R. Y Higham T. 2008. "The Red Lady" ages gracefully; new ultrafiltration AMS determinations from Paviland. *Journal of Human Evolution*, 55: 898-907.

[331] Rouillon, A. 2006. During the Gravettian, in the Cosquer cave (Marseille, Bouches-du-Rhone), has the Man counted on his fingers. *L'anthropologie* 110, 500-509.

[332] White, R. 2003. *Prehistoric Art: The symbolic Journey of humankind.* (p. 82) New York: Harry N. Abrams Inc.

[333] Svoboda, J. 2007. Upper paleolithic anthropomorph images of Northern Eurasia. In C. Renfrew & I. Morley (editors) *Image and Imagination*. Cambridge: McDonald Institute for Archaeological Research.

[334] Pettitt, P. 2002. When Burial Begins. *British Archeology* 66: 9-13.

[335] Duarte, C. Mauricio. J., Pettitt, P. Souto, P. Trinkaus, E., Plicht, H. Zilha'o, J. 1999. The early Upper Paleolithic human skeleton from the Abrigo do Lagar Velho (Portugal) and modern human emergence in Iberia. *Proceedings of the National Academy of Sciences* 96: 7604-7609.

[336] Adovasio, J.M. & Soffer, O. & Hyland, David & Illingworth, Jeff & Klíma, B. & Svoboda, Jiří. 2001. Perishable Industries from Dolní Vestonice I: New Insights into the Nature and Origin of the

Gravettian. *Archaeology Ethnology and Anthropology of Eurasia.* 2. 48-65.

[337] Klima, B. 1988. A triple burial from the Upper Paleolithic of Dolni Vestonice, Czechoslovakia, *Journal of Human Evolution* 16: 833-835.

[338] Sinitsyn, A. Hoffecker, J. 2006. Radiocarbon dating and chronology of the early upper paleolithic at Kostenki. *Quaternary International* 152-153, 175-185.

[339] Poikalainen, V. 2001. Palaeolithic art from the Danube to Lake Baikal. (Eds. M. Koiva, A. Kuperjanov, and V. Poikalainen, & E. Ernits. *Folklore, Electronic Journal of Folklore* 18 & 19: 1-60. Published by the Folk Belief and Media Group of ELM.

[340] Dimitriadis, G. 2006. From Paleolithic "Venus" up to the anthropomorphic statue-menhir. The ideological evolution of the human body in prehistoric art. *International Congress Series* 1286: 7-12.

[341] Weiss, E. Wetterstrom, W., Nadel, D., Bar-Yosef, O. 2004. The broad spectrum revisited: Evidence from plant remains. *Proceedings of the National Academy of Sciences* 101(26): 9551-9555.

[342] Larsen, C. 2005. Behavioral Reconstruction from Human Skeletal Morphology: The Last 10,000 Years Global History of Health, conference.

[343] González Sainz, César & Carmen San Miguel Llamosas. *Las cuevas del desfiladero: arte rupestre paleolítico en el valle del río Carranza (Cantabria-Vizcaya).* Santander: Ediciones Universidad de Cantabria, 2001.

[344] Brian Fagan, 2007. *People of the Earth, An Introduction to World Prehistory, Twelfth Edition.* (Section on *Painting and Engravings* p. 129.) Pearson, Prentice-Hall

[345] Susko, Michael. 2021. *The Generation of Life: Imagery, ritual & Experiences in Deep Caves.* AllrOneofUs Publishing.

[346] Wildgen, W. 2005. The paleolithic origins of art, its dynamic and topological aspects, and the transition to writing. Chapter Seven. In Bax, Marcel, Barend van Heusden and Wolfgang Wildgen (ed.). *Semiotic Evolution and the Dynamics of Culture,* Lang, Bern: 117-153.

[347] Weiss, E., Kislev, M.E., Simchoni, O., Nadel, D., & Tschauner, H. 2008. Plant-food preparation area on an Upper Paleolithic brush-hut floor at Ohalo II, Israel. *Journal of Archaeological Science,* 35(8), 2400–2414.

[348] Johnson, R. & Edgar, B. 2006. *From Lucy to Language* (p. 99) Simon and Schuster.

[349] Habu J. 2004. *Ancient Jomon of Japan.* Cambridge: Cambridge University Press.

[350] Yoshida K, Kunikita D, Miyazaki Y, Nishida Y, Miyao T, Matsuzaki H. 2013. Dating and Stable Isotope Analysis of Charred Residues on the Incipient Jomon Pottery (Japan). *Radiocarbon* 55(3):1322-1333. doi:10.1017/S0033822200048232

[351] Lee, Y. & Jong-Yoon, W. 2000. Excavation of the Sorori paleolithic site and its significance, 4th International Rice Genetics Symposium, Manilla, Philippines, Oct. 22-27, 2000.

[352] Morey, D. 2006. Burying key evidence: The social bond between dogs and people. *Journal of Archaeological Science* 33: 158-175.

[353] Bar-Yosef, O 2001. From sedentary foragers to village hierarchies: the emergence of social institutions. In *The Origin of Human Social Institutions,* ed. G. Gurnciman, 1-38. Oxford:Oxford University Press.

[354] Simons, A. 2007. *The Neolithic Revolution in the Near East* p. 46. Tuscon, AZ: The University of Arizona Press.

[355] Rice, P. C., & Moloney, N. 2008. *Biological anthropology and prehistory: Exploring our human ancestry (2nd ed.)*. Boston, MA: Allyn & Bacon (Pearson).p. 284

[356] Daniella E., Bar-Yosef, M., Porat, N. 2008. Green stone beads at the dawn of agriculture. *Proceedings of the National Academy of Sciences,*

[357] Byrd, B. 2005. Reassessing the emergence of village life in the Near East. *Journal of Archaeological Research* 13(3): 231-290.

[358] Byrd, B. 2005. Reassessing the emergence of village life in the Near East. *Journal of Archaeological Research* 13(3): 231-290.

[359] Renfrew, C. 2006. Inception of agriculture and rearing in the Middle East. *Comptes Rendus Palevol* 5: 395-404.

[360] Simmons, Alan H. 2007. *The Neolithic Revolution in the Near East: Transforming the Human Landscape.* (p. 105) Tucson: University of Arizona Press, Simmons.

[361] Peters, J. & Schmidt 2004. Animals in the symbolic world of Pre-Pottery Neolithic Gobekli Tepe, south-eastern Turkey: a preliminary assessment. *Anthropozoologica* 39(1): 179- 218.

[362] Cauvin, J. (2000). *The Birth of Gods and the Origins of Agriculture.* (p. 22). Cambridge: Cambridge University Press

[363] Simmons, Alan H. 2007. *The Neolithic Revolution in the Near East: Transforming the Human Landscape.* (p 94-95) Tucson: University of Arizona Press.

[364] Watkins, T. 2005. From foragers to complex societies in Southwest Asia. *The Human Past,* ed. Scarre, C. London: Thames and Hudson.

[365] Hayden, Brian 2003. *Shamans, Sorcerers and Saints: A Prehistory of Religions* (p. 97) Smithsonian Books.

[366] Fagan, B. 2006. *People of the Earth: Introduction to World Prehistory, 12th edition*. (p. 223.) Upper Saddle River, New Jersey: Pearson, Prentice Hall.

[367] See Transformational Stories: Voices for True Healing in Mental Health. 2019. Edited by Michael A. Susko. Baltimore: AllrOneofUs Publishing. See *Transformational Stories: Voices for True Healing in Mental Health*. 2019.

[368] Williams, J. & Taylor, E. 2006. The evolution of hyperactivity, impulsivity and cognitive diversity. *Journal Royal Society Interface* 3: 399-413.

[369] Ortega y Gasset's *Man and Crisis*, 1933 and Section on Generations Theory (pp.229-248) in John T Graham's *Theory of History in Ortega y Gasset: "The Dawn of Historical Reason."* University of Missouri Press, 1997.

[370] See *A Rosetta Key for History: Generations Revealing the Phases of Time, 2022.* Michael A. Susko.

[371] Nixon, G. (2006). Mortal Knowledge, the Orginary Event, and the Emergence of the Sacred. *Anthropoetics* 12 (1): 1-18.

Don't miss out!

Visit the website below and you can sign up to receive emails whenever Michael A. Susko publishes a new book. There's no charge and no obligation.

https://books2read.com/r/B-A-GJLJ-ZZBZE

BOOKS 2 READ

Connecting independent readers to independent writers.

Did you love *Ten Nodes of Symbolizing Humanity: From the First Hominin to the Neolithic*? Then you should read *The Logarithmic Nature of Time: From the Start of Life to Symbolizing Humanity*[1] by Michael A. Susko!

[2]

The two works collated here encompass a vast time from the start of life until the dawn to what we refer to as "civilization." In the first volume, a consistent logarithmic pulsing was found in biological evolution from life's beginning to the first humans. The evidence for these first ten nodes was primarily fossils or other life signatures in a given setting. The second volume continues the logarithmic pattering with human evolution even after the change is primarily occurring in the symbolic domain. Here, the evidence increasingly becomes geometric objects

1. https://books2read.com/u/b6oypy

2. https://books2read.com/u/b6oypy

fashioned by humans, which became tools and symbols, as well as red ochre used to decorate and signify a charged object for use in ritual.

Basically, what we mean by logarithmic patterning is that we are halving the units from the past to the future, such as 4 Gyr to 2 Gyr to 1 Gyr (billions of years) then moving on to 500 Myr, 250 Myr and on... (millions of years.) Here, a base two with exponents were used to half units and identify more refined points.

What we consistently found was evidence for increasingly complexity, an acceleration of events logarithmically, such that Major Change kept happening in less time, or more precisely half the time as the previous change. That means that change occurred in pulsed logarithmic intervals.

I have attempted to why this happens, but that is better left for another work. For now, we leave this task largely to readers to wonder about and make their own theories. We will however take on the humble and diligent task of assembling the evidence and ordering it on a logarithmic scale. Such as a scale can be updated, and I have begun some of that process. What I found was that new evidence surfacing was increasingly confirming of the logarithmic structuring outlined here.

We will let you judge yourself as to how sufficient the evidence is and what that might say about the universe. It is an exciting process, and one which can only be "in process," to be continually updated and refined by others such as yourself. Let us then discover together.

Read more at https://www.allroneofus.com/.

Also by Michael A. Susko

A Couple Through Time

Down Below & the Archon's Castle

Up Above & the Runaway

Across the Gulf & Journey Into Un-Time

On the Bay & a Child Found

Down New River & Another World

In the Wild & Do One Wild Thing

On the Mountain & Two Are Missing

To the Beginning & Journey Through Here

Archetypal Worlds

Giant Under the Mountain & the Mystery of Sacrifice

The Alien's Gift

The Gold People

Spider Woman and the Timeroc

Darkwood and Dual with the Shadow Side

Quill Ears & the Other Earth

Alwon in Another World: An Archetypal Voyage

Line In the Wall

Beyond and a New World

Beyond & a New World: I Cosmos

Beyond & A New World: II Bios

Beyond & a New World: III Mythos

Beyond & A New World: IV Logos

Biographic Book of Tens

Ten Discoveries from Biology to Spirituality: Hidden & Life-Giving Connections

Ten Times We Almost Died

Ten Sayings to Guide Our Lives

Ten Mystery Photos: Personal & Cosmological Reflections

Ten Mementos on Our Desk: Remembering Moments

Ten Metadiscoveries We Have Made

Four Years

A Survival Manual for the Next Four Years

What—Me Worry? The Next Four Years

Flirting with the Fox: The Next Four Years

New Holidays are Coming! The Next Four Years

New Monuments are Coming! The Next Four Years

New Bible Stories are Coming! The Next Four Years

New Fairy Tales Are Coming! The Next Four Years

Why I Don't Need to Repent: Four More Years!

Haikus and Photos

Flowers and Haikus
Haikus and Photos: Guatemalan Highlands
Haikus and Photos: Water Birds and Reflections
Haikus and Photos: Seasons of New River
Haikus and Photos: Yosemite Wilderness
Haikus and Photos: California Coast
Haikus and Photos: Canadian Rockies
Haikus and Photos: Hawaii's Exotic Landscapes
Haikus and Photos: Vienna: People, Buildings and Art
Haikus and Photos: Slovakian Castles and Hamlets
Haikus and Photos: Berlin, Light and Dark
Haikus and Photos: New Orleans, City of Immigrants
Haikus and Photos: Antietam Wind and Spirits
Haikus & Photos: Plant Abstractions
Haikus and Photos: Appalachian Beauty
Haikus and Photos: Urban Farm in Sandtown
Haikus and Photos: New York Heights and Ground
Haikus & Photos: Santa Fe Fractal-Pueblo Spirtuality
Haikus and Photos: Monticello's Double Vision

Little Lion

The Lion and the Chameleon
The Elephant and the Chameleons
Little Lion and His Friends
Little Lion In Flower World
Gurr in Arid Land
Little Dragon Goes to China

Logarithmic Evolutionary Time

The Logarithmic Unfolding of Astronomical Time: From Quantum Reality to Life
Ten Pulses of Evolution & the Surprising Nature of Evolutionary Time
Ten Nodes of Symbolizing Humanity: From the First Hominin to the Neolithic

Mystery Stone Explorations
What Does ChatGPT See in the Mystery Stone?

Mystery Stone Formation
Mystery Formation at Penn Bluff: Resonance with Eastern Woodland Cosmology

Nature Haikus & Photos
Haikus and Photos: Butterflies and Flowers
Haikus and Photos: A Cardinal's Life
Haikus and Photos: Two Racoons at Play
Haikus and Photos: Hawaiian Green Sea Turtle
Haikus and Photos: Irises in the Rain
Haikus and Photos: The Mystic Iris
Haikus and Photos: Tulips Opening

Paired Series of A Couple Through Time
Down Below & Up Above
Across the Gulf & On the Bay

Down New River and In the Wild
On the Mountain & To the Beginning

Rosetta Key

A Rosetta Key for History: Generations Revealing the Phases of Time
A Rosetta Key For U.S. History: Renewal or Rigidification?
A Rosetta Key for Ancestral Pueblo History: Phases of an Indigenous Civilization

Second Mystery Stone from the Shenandoah

Haikus & Photos: 2nd Shenandoan Mystery Stone
Haikus and Photos: 2nd Mystery Stone 3-D Forms

Servants of the Flower World

Flower Amid the Devastation: Servants of the Flower World, Book I
Questers in Flower World: Servants of the Flower World, Book II
Fight for This World: Servants of the Flower World, Book III
Journey to the Underworld & The Circle of Emptiness
The Edge of Return

Shenandoan Stone Explorations

Mystery Stone from the Shenandoah: Analyzed with Eastern Woodland Cosmology
Philosopher Stone from the Lower Shenandoah
Beyond the Portal: From Within the Mystery Stone
The Mystery of Essence

Gifts from the Indigenous: Another Way of Viewing & Acting in the World

Shenandoan Stone: Haikus & Photos

Haikus and Photos: Cosmogram from the Shenandoah

Haikus and Photos: Woodland Mystery Stone and World Archetypes

Haikus and Photos: Skeletal Human and Mississippian Art

Haikus and Photos: Mystery Stone's Animal Forms

Haikus and Photos: Plant Forms and Mystery Stone

Stone Formation at Penn Bluff

Haikus and Photos: Presence at Penn Bluff

Haikus & Photos: Mystery Forms at Penn Bluff

Haikus and Photos: Essences at Penn Bluff

Haikus and Photos: World Archetypes at Penn Bluff

The Dreaming Series

Sleek Back & Salamander Dreaming

Streak & Cave Bear Dreaming

Moby & Marsupial Mole Dreaming

The Dream World Trilogy

Delphi, the Time Thief, and the Dream World

Detinna and the Cave God

The Resistance & the Empire

Transformational Stories

Caseness and Narrative: Contrasting Approaches to People Psychiatrically Labeled

Transformative Experiences, Psychiatric Research, and Informed Consent

Transformational Stories: Voices for True Healing in Mental Health

Worlds to the Side

A Couple Through Time, I. From Down Below to On the Bay

A Couple Through Time, II From Down New River to Journey Through Here

Writings from Street People

Street Images

Street Images II

Standalone

Little People & the Curious Time-Riding Animal

Animal Spell: A Journey Through Darkness & Light

Child of the Elements

The Firekeeper & The Spirit of the Long Night

Life's Dynamic Vulnerability, and Its Natural Extravagance

Alien Ally

The Generation of Life: Imagery, Ritual and Experiences in Deep Caves

Twelve Suspects: And the Case of the Missing Body

2084: Clash of Cults
Bats in the Future
Guard of the Dead & Servant of the Living
The Imagination Being
Ten Traits of Empire That Every Person Should Know
Aging and Renewal: Actualizing Our LIves
The Meaning, Beauty & Mystery of Dreams: Seven Guidelines and Seven Tools for Listening
The Fragility of Evolution: Re-envisioning Life's Creative Processes
Ways We May Be Surprised by Heaven
Why Go Slow When You Can Hurl to Your Destruction?
Stages of the Human Life Cycle: A Novel Logarithmic Perspective
Fifteen Amazing Things About the Body
The Human Body as Evolutionary Treasures
What does ChatGPT see in the Mystery Photos?
Agents of Immortality: Six Ways To Living Forever
The Logarithmic Nature of Time: From the Start of Life to Symbolizing Humanity
Animal Dreaming: The Adventures of Finn
Chameleon Adventures
Dream World Trilogy
As Lover, Lion, & Phoenix
Generational Key to History: Tracing Phases From Ancient Egypt to America
Planetary Revolution: The Next Stage in Human Awareness!
Down Below to the Beginning
New Ways Are Coming! The Next Four Years
Seven Ways to Prove You're a Christian Leader
From Sunfish to Son: The Gifts of Our Lives
Cosmos & Bios: Beyond & a New World
Astronomical & Biological Unfolding of Time
Servants of the Flower World: Book I-III
Mystery Photos: Human & AI Reflections

Down Below to a New World: A Couple Through Time
Human & AI Reflections on a Mystery Stone
The Unfolding of Time: A Logarithmic Model for Astronomical, Biological, and Cultural Evolution
Servants of the Flower World: The Complete Saga I-V
Servants of the Flower World, Book IV-V: Journey to the Underworld & The Edge of Return
Generational Mapping Through Phases: Bridging Ancient Civilizations, U.S. and Ancestral Pueblo History
Moments of Sacrificial Love That the World May Live: Tales from Servants of the Flower World
Cosmos, Bios, Mythos & Logos: An Unfolding of Time Saga—in Beyond and a New World
Mythos & Symbolic Humanity: Paired Literary–Scientific Visions — Language, Stone-Shaping, and the Birth of Culture
Cosmos & Astronomical Pulses in Time: Paired Literary–Scientific Visions — Galaxies, Planets & Life
Bios & Ten Pulses of Evolution: Paired Literary–Scientific Visions — From the Origin of Life to the First Hominids
Story & Stone I: Paired Literary & Anthropological Visions: Servants of the Flower World (Books I-II) & Mystery Stone
Story & Stone II: Paired Literary & Anthropological Visions: Servants of the Flower World (Books III-V) & Ancestral Pueblo History
Epic & Opus — Paired Visions of Time: Cosmos, Bios, Mythos
A Rosetta Key for Neolithic History: Birth of Phasing—Triadic Generations at Çatalhöyük
The Birth and Unfolding of Phases in History: From Neolithic to World, Pueblo, and U.S. History
Astronomical Pulses to History's Phases: The Unfolding & Expansion of Time
A Rosetta Key for The Kingdom of Mukish: The Crossroads Civilization that Bloomed

Paired Epics, Life and Afterlife of a Couple Through Time

Watch for more at https://www.allroneofus.com/.

About the Author

With degrees in Philosophy and Psychology, the author has been dedicated to interdisciplinary and synthetic work. Previously, he published in *World Futures*, "The Fragility of Evolution," which reenvisions evolution around the concept of fragility, and protecting fragile periods of change. Since this work, he offers this addition to our evolutionary awareness, revealing its surprising logarithmic patterning.

Read more at https://www.allroneofus.com/.

AllrOneofUs Publishing

About the Publisher

AllRoneofUs Publishing seeks out work that offers a novel and qualitative contribution to world literature—the kind of work that may endure across generations.

We are especially drawn to voices that are often underheard, overlooked, or nearly lost—including authors in the later chapters of life who have made exemplary contributions that remain unrecognized. Our imprint serves as a sanctuary for works of insight, humanity, and lasting value.

Among those we have been honored to publish, we recommend:*Rich Mullin's Ethics and the Full-Breasted Richness of Life*,*John Susko's Flowers of the Night: Musings from a Sentimental Son*,and *Dr. Curtis Adams' Psychosis and the Humpty Dumpty Story*.

www.ingramcontent.com/pod-product-compliance
Lightning Source LLC
LaVergne TN
LVHW022050200926
848481LV00003B/852
9798233153327